FLORA

OF

TROPICAL EAST AFRICA

PREPARED AT THE ROYAL BOTANIC GARDENS, KEW
WITH ASSISTANCE FROM THE EAST AFRICAN HERBARIUM

EDITOR

R.M. POLHILL, B.A., Ph.D., F.L.S.

ANTHERICACEAE

BY

I. NORDAL, Fil. Dr., S. KATIVU, Ph.D.
& A.D. POULSEN, Ph.D.

CRC Press
Taylor & Francis Group
Boca Raton London New York

CRC Press is an imprint of the
Taylor & Francis Group, an **informa** business
A BALKEMA BOOK

Published by:
CRC Press/Balkema
P.O. Box 447, 2300 AK Leiden, The Netherlands
e-mail: Pub.NL@taylorandfrancis.com
www.crcpress.com – www.taylorandfrancis.com

ISBN 13: 978-90-6191-376-4 (hbk)

Sub-editor: Christopher Whitehouse.
Editorial adviser, National Museums of Kenya: G. Mwachala.
Typeset by Margaret Newman.

Visit the Taylor & Francis Web site at
http://www.taylorandfrancis.com

and the CRC Press Web site at
http://www.crcpress.com

FLORA OF TROPICAL EAST AFRICA

ANTHERICACEAE

INGER NORDAL (University of Oslo), SHAKKIE KATIVU (University of Zimbabwe) & AXEL D. POULSEN (University of Copenhagen)

Perennial herbs with rhizomes or corms, these whitish inside (not yellowish as in the closely related family Asphodelaceae). Leaves in a basal rosette, usually rosulate, sometimes subdistichous to distichous, with or without a pseudopetiole (later called petiole), linear to lanceolate, rarely terete; outer leaves often with reduced lamina (cataphylls). Peduncle usually leafless (ebracteate), sometimes with ± bract-like leaves (bracteate). Inflorescence paniculate, racemose or spicate. Flowers bisexual, hypogynous, regular to slightly irregular; tepals in two whorls of 3+3 (equal or subequal), free, white or whitish, often keeled with a coloured line of greenish, reddish or brownish; stamens 3+3, free; filaments filiform to fusiform; anthers basifixed, with longitudinally introrse dehiscence. Ovaries with several ovules per locule, developing into a loculicidal capsule. Seeds flat or folded, rarely turgid, black, ± glossy, without an aril (not dull and arillate as in Asphodelaceae).

A tropical to temperate family with 8 genera, of which 2 occur in Africa; the number of species is difficult to estimate due to uncertain delimitation at different levels. The family is closely related to Asphodelaceae, from which it may sometimes be difficult to distinguish by its gross morphology. Representatives of Asphodelaceae, however, always contain anthraquinones, displayed especially by the rhizomes which are yellowish when cut; in Anthericaceae the rhizome is always whitish inside. The seeds of Asphodelaceae are covered by a thin aril, making them brownish to greyish and dull; in Anthericaceae the seeds are always black, sometimes glossy.

Roots ± swollen, without tubers; flowers solitary at each node of the inflorescence, supported by only one bract; pedicels without a joint; seeds ± turgid · · · · · · · · · · · · · · · 1. **Anthericum**

Roots swollen or, if not, bearing tubers; flowers usually more than one at each node of the inflorescence or, if only one then supported by two bracts; pedicels generally articulated; seeds ± thin, flat or folded · · · · · · · · · · · · · · · · 2. **Chlorophytum**

1. ANTHERICUM

L., Sp. Pl.: 310 (1753); Baker in F.T.A. 7: 477 (1898), pro parte; Oberm. in Bothalia 7: 669–767 (1962), pro parte; Nordal et al. in Mitt. Inst. Allg. Bot. Hamburg 23: 535–559 (1990); Kativu & Nordal in Nordic Journ. Bot. 13: 59–65 (1993); Nordal & Thulin in Nordic Journ. Bot. 13: 257–280 (1993)

Small plants, up to 20 cm. tall. Rhizome very short; roots ± swollen, without tubers. Cataphylls membranous, often hidden by fibrous remnants from previous years' leaves. Leaves in a basal rosette, linear to ± succulent. Peduncles leafless, often several to a plant, reduced, within the leaf rosette, or elongated, up to 15 cm. Inflorescence umbel-like (pedicels then emerging from inside the leaf rosette) or a simple raceme; pedicels 1 per node, subtended by a single bract, never articulated.

Perianth persistent as fruit matures. Tepals white, often with greenish tinge or stripe outside. Filaments filiform, smooth. Capsule rounded or shallowly 3-lobed in cross-section. Seeds turgid, only slightly folded.

A mainly Old World temperate genus with ± 10 species.

1. Peduncle completely reduced; inflorescence umbel-like, the pedicels 3–8 cm. long, emerging directly from the leaf rosette · · · · · · · · · · 1. *A. angustifolium*
   Peduncle distinct, up to 15 cm. long; inflorescence racemose, unbranched or with few basal branches, pedicels shorter than 2.5 cm. · · · · · · · · · · 2
2. Leaves ± 0.5 cm. wide, semi-succulent to fistulose; inflorescence never branched; capsule slightly ridged, but not verrucose · · · · · · · · · · 2. *A. corymbosum*
   Leaves ± 1 cm. wide, flat; inflorescence often with 1–3 basal branches; capsule verrucose · · · · · · · · · · 3. *A. jamesii*

1. **A. angustifolium** *A. Rich.*, Tent. Fl. Abyss. 2: 331 (1850); Baker in F.T.A. 7: 489 (1898); W.F.K.: 117 (1948); Polhill in Journ. E. Afr. Nat. Hist. Soc. 24: 6 (1962); E.P.A.: 1530 (1971); Blundell, Wild Fl . E. Afr., reprint: 418 (1992); Nordal & Thulin in Nordic Journ. Bot. 13: 258 (1993); U.K.W.F., ed. 2: 316, t. 144 (1994); Nordal in Fl. Ethiopia 6: 90 (1997). Lectotype, chosen by Nordal & Thulin (1993): Ethiopia, Tigray, Chiré, *Quartin-Dillon* (P, lecto.!, K, isolecto.!)

Small plants up to ± 15 cm. Leaves linear, 5–15 cm. long, 2–5 mm. wide, glabrous or with ciliate margins. Peduncle very short and hidden among the leaves; rhachis almost completely reduced, making the inflorescence umbel-like. Pedicels 3–8 cm. long. Flowers white, greenish tinged outside, star-like, 2–10 to a plant; tepals 8–12 mm. long, 2–3 mm. wide, 3-veined. Stamens shorter than the tepals, with filaments up to 4 mm. long; the yellow anthers shorter and curled after anthesis. Capsule subglobose, 5–8 mm. long, smooth or slightly ridged, on lax pedicels so that they end up lying on the ground. Seeds ± 2 mm. in diameter.

UGANDA. Mbale District: Sebei, Kaburoron [Kabururon], 15 May 1955, *Norman* 259!
KENYA. Northern Frontier Province: Lorogi Plateau, Kisima, 14 Apr. 1977, *Timberlake* 1034!; SW. Elgon, 11 May 1958, *Symes* 358!; Nakuru District: Ndoinet R. valley, 28 Mar. 1970, *Gillett* 19025!; Londiani, near forest office, 13 May 1949, *Maas Geesteranus* 4650!
DISTR. **U** 3; **K** 1, 3, 5; Eritrea and Ethiopia
HAB. Growing in clumps in upland grassland, seasonally waterlogged, often heavily grazed and eroded, on shallow and gritty black or lateritic soils, sometimes in crevices in the rock; 1800–2850 m.

SYN. *A. humile* A. Rich., Tent. Fl. Abyss. 2: 331 (1850); Baker in F.T.A. 7: 489 (1898); E.P.A.: 1531 (1971). Lectotype, chosen by Cufod. (1971): Ethiopia, Tigray, Oudgerate, *Petit* (P, lecto.!, K, isolecto.!)

2. **A. corymbosum** *Baker* in J.B. 15: 71 (1877) & in F.T.A. 7: 489 (1898); W.F.K.: 117 (1948); E.P.A.: 1530 (1971); Nordal & Thulin in Nordic Journ. Bot. 13: 258 (1993); U.K.W.F., ed. 2: 316 (1994); Thulin, Fl. Somalia 4: 42, fig. 26 (1995); Nordal in Fl. Ethiopia 6: 91 (1997). Type: Somalia, Meid, Mt. Surrut, *Hildebrandt* 1471 (K, holo.!)

Plants 10–20 cm. Leaves linear, 5–25 cm. long, 1–8 mm. wide, ± succulent, often with ciliate margins. Peduncles 3–15 cm. long, glabrous. Inflorescence a simple raceme, 3–7 cm. long, with 3–15 flowers; floral bracts membranous, cuspidate, up to 12 mm. long. Pedicels semipatent, sometimes drooping in fruit, 5–25 mm. long, elongating to 40 mm. with age. Tepals white with green stripes on the outside, 9–10

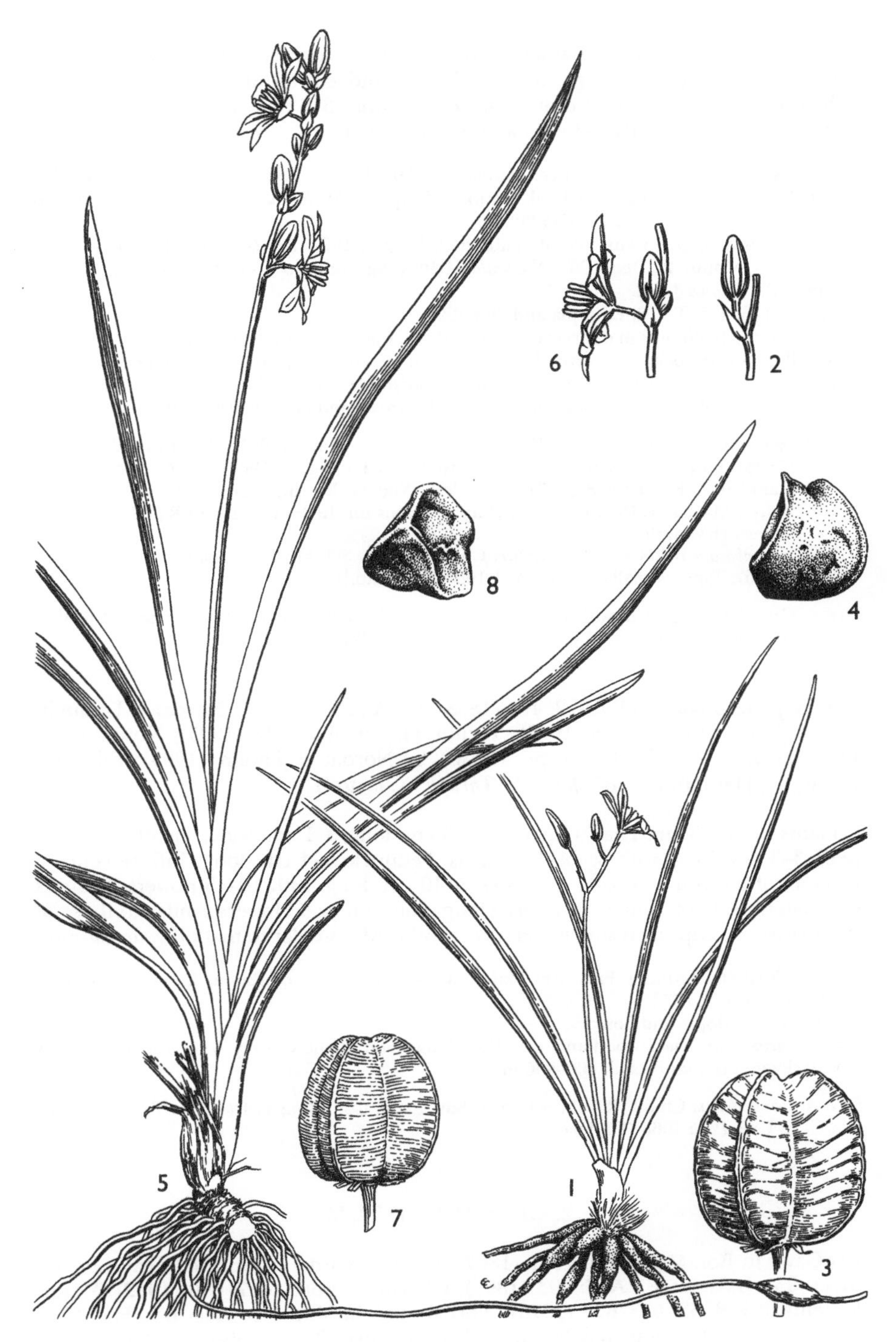

FIG. 1. *ANTHERICUM CORYMBOSUM* — **1**, habit, × ²⁄₃; **2**, node of inflorescence, × 1; **3**, capsule, × 2; **4**, seed, × 8. *CHLOROPHYTUM CAMERONII* var. *PTEROCAULON* — **5**, habit, × ²⁄₃; **6**, node of inflorescence, × 1; **7**, capsule, × 2; **8**, seed, × 8. 1, 2, from *Mathew* 6039; 3, 4, from *Polhill* 372A; 5, 6, from *Mathew & Hanid* 6050; 7, 8, from *Bidgood, Mwasumbi & Vollesen* 1226. Drawn by Eleanor Catherine.

mm. long, 2–3 mm. wide. Stamens shorter than the tepals, with filaments up to 4 mm. long; the yellow-orange anthers shorter and curled after anthesis. Capsule subglobose with a rounded triangular cross-section, 3–8 mm. long, slightly ridged, but not verrucose. Seeds 2–3 mm. across. Fig. 1/1–4.

KENYA. Northern Frontier Province: Moyale, 17 Apr. 1952, *Gillett* 12834!; Naivasha District: Gilgil to Naivasha road, near Ilkek Station, 13 Apr. 1970, *Mathew* 6039!; Nairobi National Park, 22 Apr. 1973, *Gillett & White* 20233!

TANZANIA. Serengeti, 15 km. SW. of Haibardad, 17 Dec. 1973, *Kreulen* 207!; Masai District: S. Serengeti, Mpitin, 22 Dec. 1962, *Newbould* 6429! & Ngorongoro Conservation Area, Esere, 3 Dec.1989, *Chuwa* 3061!

DISTR. **K** 1, 3, 4, 6; **T** 1, 2; Ethiopia and Somalia

HAB. Growing in clumps in depressions in grassland, bushland and degraded *Acacia-Combretum* woodland or on rocky outcrops; often on shallow soils overlying rock (often lava, sometimes limestone), seasonally waterlogged, on heavy clayish black or red lateritic soils or on lighter brownish to reddish sandy soils; often in heavily grazed and eroded areas; 400–2900 m.

SYN. *A. gregorianum* Rendle in J.L.S. 30: 416 (1895); P.O.A. C: 139 (1895); Baker in F.T.A. 7: 490 (1898); Polhill in Journ. E. Afr. Nat. Hist. Soc. 24: 6, fig. 3 (1962); E.P.A.: 1530 (1971); Hanid in U.K.W.F.: 678 (1974); Blundell, Wild Fl. E. Afr., reprint: 419 (1992). Type: Kenya, Machakos District, Kapiti [Kapte] Plains on the banks of Athi R., 24 Apr. 1893, *Gregory* (BM, holo.!)

A. corymbosum Baker var. *floribundum* Chiov., Result. Sci. Miss. Stef.-Paoli, Coll. Bot.: 173 (1916). Type: Somalia, Uegit, *Paoli* 1084 (FT, lecto.!)

NOTE. One deviating form with reflexed pedicels and small fruits has been collected in **K** 1, Mandera District, 6 km. S. of El Wak (*M.G. Gilbert & Thulin* 1230!).

3. **A. jamesii** *Baker* in F.T.A. 7: 490 (1898); E.P.A.: 1531 (1971); Nordal & Thulin in Nordic Journ. Bot. 13: 258 (1993); Thulin, Fl. Somalia 4: 42 (1995); Nordal in Fl. Ethiopia 6: 91 (1997). Lectotype, chosen by Nordal & Thulin (1993): Ethiopia, Hararghe, Hahi, June 1885, *James & Thrupp* (K, lecto.!)

Plants up to 15 cm. Leaves linear, 10–15 cm. long, ± 1 cm. wide, sheathing into a neck 3–5 cm. long, with scabrid margins. Peduncle 1–4 cm. long. Inflorescence a raceme, often with 1–3 basal branches, 5–10 cm. long, with 10–15 flowers. Pedicels semipatent, ± 1 cm. long at anthesis, elongating with age. Flowers similar to those of *A. corymbosum*. Capsule distinctly verrucose, otherwise similar to that of *A. corymbosum*.

KENYA. Northern Frontier Province: Mandera, 5–10 km. SSE. of Ramu, 2 May 1978, *M.G. Gilbert & Thulin* 1333!

DISTR. **K** 1; Ethiopia and Somalia

HAB. *Acacia-Commiphora* woodland on alluvial flats, on shallow stony soils, on temporarily waterlogged red sandy soil; 400–900 m.

SYN. *A. verruciferum* Chiov., Result. Sci. Miss. Stef.-Paoli, Coll. Bot.: 174 (1916). Type: Somalia, El Uré, *Paoli* 1068 (FT, lecto.!)

2. CHLOROPHYTUM

Ker-Gawl. in Bot. Mag. 27, t. 1071 (1807), emend. Nordal in Nordic Journ. Bot. 13: 61 (1993); Baker in F.T.A. 7: 493 (1898); Oberm. in Bothalia 7: 690 (1962); Marais & Reilly in K.B. 32: 653–663 (1978); Nordal et al. in Mitt. Inst. Allg. Bot. Hamburg 23: 535–559 (1990); Kativu & Nordal in Nordic Journ. Bot. 13: 59–65 (1993); Nordal & Thulin in Nordic Journ. Bot. 13: 257–280 (1993); Stedje & Nordal in Seyani & Chikuni, Proc. XIII AETFAT, Malawi, 1991: 513–524 (1994)

Dasystachys Baker in Trans. Linn. Soc., Bot., ser. 2, 1: 255 (1878) & in F.T.A. 7: 510 (1898); E.P.A.: 1537 (1971), *nom. illegit.*, *non* Oersted (1859)

Acrospira Baker in Trans. Linn. Soc., Bot., ser. 2, 1: 255 (1878) & in F.T.A. 7: 477 (1898), *nom. illegit.*, *non* Mont. (1857), *nec* Berk & Broome (1861)
Debesia Kuntze, Rev. Gen. Pl. 2: 708 (1891), *nom. nov.* for *Acrospira* Baker
[*Anthericum* sensu Baker in F.T.A. 7: 477 (1898), pro parte; Oberm. in Bothalia 7: 670 (1962), pro parte; Hanid in U.K.W.F.: 674 (1974), pro parte, *vide* Kativu & Nordal in Nordic Journ. Bot. 13: 59–65 (1993)]
Verdickia De Wild. in Ann. Mus. Congo, Bot., sér. 4, 1: 7 (1902)

Plants very variable in size and robustness, from 5 cm. to more than 1 m. Rhizome reduced or prominent, sometimes moniliform (that is consisting of a series of attached corms in a chain); roots either swollen without tubers or ± wiry with distinct tubers. Leaves in a basal rosette, rosulate or distichous, linear to broadly lanceolate, sometimes narrowed towards the base to a petiole. Peduncles reduced or elongated up to 50 cm., leafless or with ± bract-like leaves. Inflorescence spicate, racemose or paniculate, with multibracteate nodes. Pedicels, with very few exceptions, with a distinct joint (articulated), sometimes only one at each node, but in most species several (multi-nate). Flowers open, star-shaped, urceolate to almost bell-shaped. Tepals white to yellowish green, with or without greenish or reddish stripes on the outside. Stamens straight or declinate; filaments filiform or fusiform, glabrous or papillate. Ovary sessile, rarely stipitate; style declinate or straight. Capsule, ± triangular (trigonous) to deeply 3-lobed (triquetrous) in cross-section. Seeds thin, flat or sharply folded.

An Old World tropical to subtropical genus with ± 150 species and centre of diversity in tropical Africa.

About half the species in East Africa belong to three distinctive groups of species:
Species 8–13 includes related taxa with spongy roots without tubers, spicate inflorescences and bell-shaped flowers. In Floras and revisions up to ± 1970 they were often referred to a separate genus *Dasystachys*.

Species 22–26 includes closely related taxa, with distichous leaves and characteristic roots, flowers and fruits. They represent the core of the tropical genus '*Anthericum*' as it was delimited by all tropical African Floras and revisions before 1990. These taxa are, however, more closely related to other species of *Chlorophytum* than they are to the mainly temperate genus *Anthericum* as delimited here (cf. Nordal et al. in Mitt. Inst. Allg. Bot. Hamburg 23 (1990) and Kativu & Nordal in Nordic Journ. Bot. 13 (1993)).

Species 32–49 includes related taxa, all of which have a basic chromosome number of x=7 as far as known, thus differing from the rest of the genus and family where x=8 appears to be the rule. The x=7 group is probably monophyletic and derived. The type species of the genus *Chlorophytum*, the West African *C. inornatum* Ker-Gawl. also belongs in this group. The taxa within this group will probably often hybridize when they meet, and specific delimitation might be difficult in several cases. Species 35–39 constitute a cluster of species with mainly woodland ecology, characterized by having tubers on lateral branches and with a tendency to dry brownish; whereas species 46–48 have a clear preference for forests and forest edges, characterized by having tubers along the main root-axes and with a tendency to dry blackish. Species 43–45 are reduced forms which may have developed from larger forms of either of the two groups.

1. Leaves along the entire peduncle · · · · · · · · · · 2
   Leaves only in a basal rosette (occasionally some sterile bracts below the inflorescence) · · · · · · · · · · 11
2. Inflorescence richly branched (paniculate) · · · · · · · · · · 3
   Inflorescence rarely branched (racemose to spicate), only occasionally with a few basal branches (when *C. ducis-aprutii* is cultivated it becomes branched) · · · · · · · · · · 7
3. Basal leaves reduced; leaves on peduncle dominant, clasping, large and leafy · · · · · · · · · · 4. *C. ruahense*
   Basal leaves not reduced; leaves on peduncle and bracts narrow, neither dominant nor clasping · · · · · · · · · · 4
4. Roots not spongy with elongated tubers distally; leaves pubescent · · · · · · · · · · 51. *C. sp. A*
   Roots spongy without tubers; leaves glabrous · · · · · · · · · · 5

5. Basal leaves petiolate, broadly lanceolate 3–9 cm. wide; tepals greenish, (9–)12–20 mm. long, with a constriction in lower part, ligulate above the constriction; ovary stipitate · · · · · · · · · · · · · · · · · · 1. *C. andongense*
   Basal leaves not petiolate, narrowly lanceolate, up to 2.5 cm. wide; tepals whitish, 6–11 mm. long, without a constriction and not ligulate; ovary sessile · · · · · · · · · · · · 6
6. Leaf margin undulate; perianth urceolate; tepals reflexed; filaments fusiform · · · · · · · · · · · · · · · · · 2. *C. viridescens*
   Leaf margin not undulate; perianth open, ± stellate; tepals not reflexed; filaments filiform · · · · · · · · · · · 3. *C. nyasae*
7. Inflorescence a dense spike; pedicels articulated at the apex; perianth bell-shaped; tepals covering the ovary at anthesis · 8
   Inflorescence a ± open raceme; pedicels articulated near or below the middle; perianth open, stellate; tepals not covering the ovary at anthesis · · · · · · · · · · · · 9
8. Peduncle densely papillate-pubescent above; flowers 2 or more at each node, tepals with 3 veins; leaf-bases without purple spots · 8. *C. longifolium*
   Peduncle glabrous to subpapillose; flowers only 1 at each node, tepals with 1 vein; leaf-bases with purple spots · 9. *C. colubrinum*
9. Plants usually less than 50 cm.; leaves produced after the flowers (hysteranthous); roots without tubers · · · · · 6. *C. nubicum*
   Plants 60–250 cm.; leaves produced with the flowers (synanthous) · 10
10. Capsules smooth, deeply lobed in cross-section (triquetrous); seeds ± flat; roots spongy without elongated tubers; plants from N. Kenya · · · · · · · · · 7. *C. ducis-aprutii*
    Capsules with transverse ridges, rounded in cross-section (trigonous); seeds irregularly folded; roots with elongated tubers; plants from SW. Tanzania · · 5. *C. stolzii*
11. Only one flower/fruit in the nodes of the inflorescence (in *C. bifolium* rarely 2 at the lower node) · · · · · · · · · · · · 12
    Two or more flowers/fruits in the lower nodes of the inflorescence · 24
12. Plants with a distinct aerial stem, up to 50 cm. long, with ring-like leaf scars · 28. *C. suffruticosum*
    Plants with subterranean rhizomes, never distinct aerial stems · 13
13. Roots spongy, ± tuber-like; pedicels articulated just below the flower; flowers bell-shaped, tepals ± erect, papillate inside in a zone just above the ovary · · · · · · · · · 14
    Roots spongy or not, with or without delimited tubers; pedicels distinctly articulated at a distance below the apex (if apically, then a small plant with tuber-like roots, i.e. *C. collinum*); flowers open, stellate or zygomorphic, tepals patent or ± reflexed without papillation · 17
14. Plants more than 10 cm. high; leaves erect, glabrous or ciliate only along the margins; peduncle ± glabrous; capsules longer than 2 mm. · 15
    Plants up to 10(–12) cm. high; leaves ± prostrate, pubescent, peduncle pubescent; capsules up to 2 mm. long · 16

15. Leaves up to 1.5 cm. wide, margins glabrous or nearly so (rarely white-ciliate); floral bracts not awned, glabrous or nearly so .... 10. *C. silvaticum*
    Leaves wider than 1.5 cm., margins red-ciliate; floral bracts awned, margins red-ciliate .... 11. *C. africanum*
16. Rhizome short, moniliform, never stoloniferous and elongated, cataphylls and peduncles with purple coloration .... 12. *C. macrorrhizum*
    Rhizome long, stoloniferous, never moniliform; cataphylls and peduncles without purple coloration 13. *C. leptoneurum*
17. Flowers zygomorphic with tepals longer than 15 mm.; capsule longer than 10 mm. .... 19. *C. somaliense*
    Flowers regular with tepals shorter than 12 mm. long; capsule shorter than 10 mm. 18
18. Pedicels articulated at the apex; tepals with one vein 14. *C. collinum*
    Pedicels articulated near or below the middle; tepals with 3 veins 19
19. Leaves distichous; tepals longer than 6 mm. .... 15. *C. affine*
    Leaves rosulate; tepals up to 6 mm. long 20
20. Roots ± fleshy, tuber-like or with elongated tubers distally; capsules slightly lobed in cross-section (trigonous) 21
    Roots thin and wiry with distinctly delimited, ± globose tubers; capsules deeply lobed in cross-section (triquetrous) 23
21. Roots tuber-like; rhachis pubescent; filaments filiform 18. *C. bifolium*
    Roots with elongated tubers distally, rhachis glabrous; filaments fusiform 22
22. Leaves long, linear and grass-like, up to 0.5 cm. wide, shortly ciliate along the apical parts; inflorescence a lax raceme, 3–10 cm. long; in eastern coastal areas 49. *C. tenerrimum*
    Leaves short, broadly lanceolate, ± 4 cm. wide, completely glabrous; inflorescence a condensed raceme, up to 2 cm. long; in western rain forest areas 52. *C. sp. B*
23. Leaves and rhachis glabrous, leaves filiform; peduncle reduced, so that flowers and fruits appear among the leaf-bases; capsules 8–10 mm. long .... 30. *C. inconspicuum*
    Leaves and rhachis pubescent, leaves narrowly lanceolate; peduncle well developed, so that flowers and fruits are exposed; capsules ± 3 mm. long .... 17. *C. fischeri*
24. Pedicels apparently not articulated or articulated at the apex; peduncle stout, less than 3 cm. long; inflorescence condensed 25
    Pedicels distinctly articulated, peduncle longer than 5 cm. long, if shorter, then plants with ± prostrate, lax inflorescences 28
25. Plants up to 10 cm. high, not drying blackish; leaves ± prostrate, broadly oblanceolate, less than 20 cm. long 26
    Plants more than 15 cm. high, drying blackish; leaves suberect, lanceolate, longer than 20 cm. 27
26. Roots fibrous, bearing tubers at the tips; leaves petiolate; pedicels longer than 4 mm. long, recurved in fruit stage; tepals with 5 veins .... 44. *C. geophilum*
    Roots spongy, often swollen distally; leaves not petiolate; pedicels up to 3 mm. long, not recurved; tepals with 3 veins .... 45. *C. pusillum*

27. Leaves lanceolate, up to 4 cm. wide, without a petiole; perianth greenish .... 43. *C. stenopetalum*
    Leaves broadly lanceolate, more than 5 cm. wide, with a distinct petiole; perianth whitish .... 47. *C. filipendulum* subsp. *amaniense*
28. Leaves distichous (except in *C. paucinervatum*); buds and flowers tinged/streaked purplish to pinkish; peduncles often flat and winged or ribbed; capsules transversely ridged; seeds irregularly folded 29
    Leaves rosulate or distichous; buds and flowers tinged/streaked greenish (rarely brownish); peduncles ± terete, not winged/ribbed; capsules without distinct transverse ridges; seeds usually flat to saucer-shaped 33
29. Cataphylls, leaves and bracts with purple or pink coloration; perianth pinkish; mature capsules with purple coloration .... 26. *C. rubribracteatum*
    Cataphylls and leaves green (leaf-bases sometimes with purple or red stripes in *C. cameronii*); perianth white with a pinkish tinge or tip; capsules green 30
30. Outer tepals with red-black tips; stamens arranged with radial symmetry .... 25. *C. sphacelatum*
    Outer tepals tinged pink on the outside; stamens usually in 2+4 arrangement 31
31. Leaves lanceolate, wider than 5 mm., rarely ciliate/pubescent .... 22. *C. cameronii*
    Leaves filiform to linear, narrower than 4 mm., ciliate 32
32. Leaves (sub)distichous, filiform, up to 2 mm. wide, without red-brownish bands at the base; tepals up to 10 mm. long .... 23. *C. goetzei*
    Leaves rosulate, linear, grass-like, 2–4 mm. wide, with prominent red-brownish bands at the base; tepals longer than 10 mm. .... 24. *C. paucinervatum*
33. Roots with large distinct tubers, longer than 5 cm.; tepals 6–9 mm. wide, with 9 or more veins .... 20. *C. tuberosum*
    Roots without or with smaller tubers; tepals less than 6 mm. wide, with 1–5(–7) veins 34
34. Peduncle shorter than 3 cm.; peduncle and inflorescence ± prostrate 35
    Peduncle longer than 5 cm., erect (sometimes arcuate below) 38
35. Leaves distichous, more than 5 mm. wide; rhachis densely papillate to pubescent 36
    Leaves rosulate, less than 5 mm. wide; rhachis ± glabrous 37
36. Roots with tubers on lateral branches; leaves glabrous, not blotched with spots basally; tepals 5–7 mm. long 37. *C. humifusum*
    Roots with tubers along the main axes; leaves ciliate, blotched with greenish/brownish spots basally; tepals 8–11 mm. long .... 15. *C. affine* var. *curviscapum*
37. Inflorescence paniculate; tepals 5–8 mm. long; capsule subglobose, shallowly lobed, less than 5 mm. long 29. *C. angustissimum*
    Inflorescence simple; tepals 3–5 mm. long; capsule deeply lobed, more than 6 mm. long .... 30. *C. inconspicuum*

38. Pedicels stiffly patent, longer than 20 mm.; plants from N. Kenya ··· 21. *C. zavattari*
Pedicels not stiffly patent, shorter than 20 mm. ··· 39
39. Rhizome distinctly moniliform with roots thick near the base, gradually tapering (fusiform), never with tubers; capsules rounded in cross-section (trigonous); seeds 1.5 mm. in diameter, irregularly folded ··· 27. *C. subpetiolatum*
Rhizome variable, roots never fusiform, with ± prominent tubers; capsules ± deeply lobed in cross-section (triquetrous); seeds 2 mm. or more in diameter, flat to saucer-shaped ··· 40
40. Plants with ± condensed, unbranched inflorescences; tepals 8–15 mm. long, 5-veined; anthers 4–6 mm. and longer than the filaments ··· 39. *C. macrophyllum*
Plants with paniculate to unbranched, lax to moderately condensed inflorescence; tepals usually shorter than 8 mm. (if longer, then plants with distichous leaves, *C. affine*), usually 1–3 veined; anthers less than 5 mm. and shorter than the filaments ··· 41
41. Leaves crisped and ciliate at the margin; ovary and capsule papillate to tuberculate ··· 42. *C. brachystachyum*
Leaves never crisped; ovary and capsules ± smooth ··· 42
42. Cataphylls prominent and ciliate; plants drying blackish; inflorescence racemose, condensed, rarely with a few basal branches ··· 43
Cataphylls not prominent nor ciliate; plants drying blackish or greenish; inflorescence racemose to paniculate, not particularly condensed ··· 44
43. Leaves narrowly lanceolate, more than 4 times longer than wide ··· 40. *C. blepharophyllum*
Leaves broadly lanceolate, only 2–3 times longer than wide ··· 41. *C. amplexicaule*
44. Grassland to woodland species (except *C. holstii*, which is found in forest); never drying blackish; roots fibrous to moderately swollen, never tomentose ··· 45
Forest to riverine species; usually drying blackish; roots often swollen, tomentose; flowers whitish ··· 56
45. Inflorescence racemose; flowers whitish ··· 46
Inflorescence paniculate; flowers greenish ··· 50
46. Leaves distichous; peduncle angled; tepals longer than 8 mm. ··· 15. *C. affine*
Leaves rosulate; peduncle terete; tepals shorter than 7 mm. ··· 47
47. Pedicels slender, drooping (except in early flowering stage); capsules pyramid shaped in outline, 10–15 mm. long, hanging ··· 16. *C. pendulum*
Pedicels never drooping; capsules ovoid, shorter than 7 mm. long, erect ··· 48
48. Roots with tubers on lateral branches; leaves distinctly petiolate, tepals longer than 5 mm. ··· 48. *C. holstii*
Roots with tubers along the main axes; leaves not petiolate, tepals usually shorter than 5 mm. ··· 49
49. Peduncle and rhachis papillate to pubescent, stiffly erect 17. *C. fischeri*
Peduncle and rhachis glabrous, lax ··· 50. *C. tenerrimum*

50. Roots with tubers on lateral branches; leaves rosulate, never with brownish-green bands at the bases; plants glabrous or slightly papillose in upper parts 51
    Roots with tubers along the main axes; leaves more or less distichous; leaves, peduncles and/or flowers ciliate or pubescent (except *C. polystachyum* which may be glabrous, but then with broad brownish-green bands at the leaf-bases) 54
51. Leaves petiolate; bracts large and leafy, pedicels angled to winged above the articulation . . . . 38. *C. zingiberastrum*
    Leaves not petiolate; bracts small, pedicels not angled nor winged above the articulation 52
52. Upper part of peduncle, rhachis and pedicels papillate; capsules longer than broad, ± 7 mm. long . . . . 36. *C. floribundum*
    Peduncle, rhachis and pedicels glabrous; capsules broader than long, less than 4 mm. long 53
53. Leaves contemporary with flowers, broadly lanceolate, more than 2 cm. wide; tepals 5–7 mm. long . . . . . . 34. *C. gallabatense*
    Leaves hysteranthous, narrowly lanceolate, less than 2 cm. wide; tepals 4–5 mm. long . . . . 35. *C. micranthum*
54. Peduncle pilose; leaves pubescent, without a brownish-green band at the base, capsule up to 4 mm. long 33. *C. vestitum*
    Peduncle glabrous; leaves glabrous or ciliate along the margins (rarely also on the veins), often with a broad brownish-green band at the base; capsules 4–8 mm. long 55
55. Pedicels and perianth glabrous . . . . 31. *C. polystachys*
    Pedicels and perianth papillate to pubescent . . . . 32. *C. pubiflorum*
56. Robust plants, always drying black; leaves petiolate usually broader than 5 cm.; rhachis distinctly scabrid; capsules more than 7 mm. long, usually distinctly longer than wide . . . . 47. *C. filipendulum* subsp. *filipendulum*
    Slender plants, sometimes drying black; leaves petiolate or not, narrower than 5 cm.; rhachis glabrous to slightly scabrid capsules usually less than 7 mm. long and wider than long (except in a mountain form of *C. comosum*) 57
57. Leaves petiolate or not, petioles shorter than the lamina; lamina narrowly lanceolate; base attenuate; inflorescence often with small plantlets (pseudovivipary) . . . . 46. *C. comosum*
    Leaves distinctly petiolate, petioles longer than the lamina; lamina broadly lanceolate; base truncate to cordate; never with pseudovivipary . . . . 48. *C. lancifolium*

1. **C. andongense** *Baker* in Trans. Linn. Soc., Bot., ser. 2, 1: 260 (1878); P.O.A. C: 140 (1895); Baker in F.T.A. 7: 506 (1898); F.P.S. 3: 267 (1956); Polhill in Journ. E. Afr. Nat. Hist. Soc. 24: 12 (1962); Hepper in F.W.T.A., ed. 2, 3: 102 (1968). Type: Angola, Pungo Andongo, *Welwitsch* 3770 (BM, holo.!, K, iso.!)

Robust plants, 65–200 cm. high. Rhizome short, thick, vertical, moniliform, bearing fibrous remains of old leaf-bases; roots spongy, fusiform, extensive, without tubers. Leaves rosulate, sheathing at the base, oblong-lanceolate, petiolate, glabrous, 25–80 cm. long, 3–8 cm. wide; margins not undulate. Peduncle terete, glabrous, 5–7 mm. in diameter, with deciduous leaves all along. Inflorescence a lax panicle, much exserted above the leaves, up to 80 cm. long; flowers sometimes congested in the apical parts of the branches; floral bracts lanceolate, glabrous. Pedicels articulated above the middle, up to 30 mm. long in fruit, 4–9-nate at the lower nodes. Flowers ± pendulous, often congested; perianth greenish to whitish, urceolate; tepals (9–)12–20 mm. long, 2–3 mm. wide, with ligulate rims above the constriction to the cup, 3-veined, scabrid on the margins. Stamens shorter than the perianth; filaments fusiform, with transverse papillate ridges, longer than the anthers. Ovary stipitate; style nearly as long as the stamens. Capsule obovoid, nearly rounded in cross-section, 10–12 mm. long. Seeds disc-shaped, 3 mm. in diameter. Fig. 2/1–2.

UGANDA. W. Nile District: Arua, 16 Jan. 1945, *Greenway & Eggeling* 7345!; Acholi District: Paloga, Apr. 1943, *Purseglove* 1357!; Karamoja District: Amudat, Nov. 1964, *Tweedie* 2939!

KENYA. Uasin Gishu District: hills above Moiben R., Dec. 1931, *Brodhurst Hill* 671!; Ravine District: Kampi ya Moto–Eldama, 11.5 km. from Eldama Ravine, 25 Sept. 1953, *Drummond & Hemsley* 4435!; Teita District: Mt. Kasigau, 26 Apr. 1963, *Bally* 12699A!

TANZANIA. Masai District: Ngorongoro Crater, Apr. 1941, *Bally* 2400!; Pare District: Lembeni, 30 July 1957, *Bally* 11619!; Songea District: Kwamponjore valley, 7 Feb.1956, *Milne-Redhead & Taylor* 8712!

DISTR. **U** 1, 3; **K** 1, 3, 6, 7; **T** 1–8; Guinea, Sierra Leone, Nigeria, Sudan, Angola, Zambia, Malawi, Mozambique and Zimbabwe

HAB. In open woodland, forest margins and riverine thicket, often on termite mounds; 150–2100 m.

SYN. *C. longipes* Baker in J.B. 16: 325 (1878). Types: Sudan, Jur [Djur], *Schweinfurth* 1801 & 2045 (K, syn.!)

Anthericum gossweileri Poelln. in Bol. Soc. Brot., sér. 2, 17: 67 (1943). Type: Angola, Malange, *Gossweiler* 943 (B, holo.!)

Chlorophytum pedunculosum Poelln. in Portug. Acta Biol., sér. B, 1: 323 (1946). Type: Tanzania, Kigoma District, Uvinza [Uvinsa], SW. of Malagarasi, *Peter* 35874 (B, holo.!)

C. viridescens Engl. var. *majus* Poelln. in Portug. Acta. Biol., sér. B, 1: 350 (1946). Type: Tanzania, Dodoma District, Ugogo, between Kitalalo and Chali [Tschali], *Peter* 33310 (B, holo.!)

NOTE. Material from East Africa has sometimes been misidentified to *C. macrosporum* Baker, a related South African species so far known N. to Zimbabwe and Mozambique. The latter is characterized by narrower leaves without petioles and larger (11–14 mm. long) and more deeply lobed (triquetrous) capsules. Hanid in U.K.W.F. (1974) reported the species from Kenya probably due to wrong identification. The material he referred to either belongs in this species or the next, *C. viridescens*. The two species might be difficult to delimit.

2. **C. viridescens** *Engl.*, P.O.A. C: 140 (1895); Baker in F.T.A. 7: 509 (1898); Polhill in Journ. E. Afr. Nat. Hist. Soc. 24: 15 (1962); U.K.W.F., ed. 2: 317 (1994). Type: Tanzania, Kilimanjaro, below Marangu, *Volkens* 2155 (B, holo.!, K, iso.!)

Plants up to 120 cm. high. Rhizome vertical, short, thick, bearing fibrous remains of old leaf-bases; roots thick (diameter up to 1 cm.) and spongy, whitish, extensive, without tubers. Leaves often hysteranthous, rosulate, linear to narrowly lanceolate, not petiolate, glabrous, 20–45 cm. long, (0.6–)1–2.5 cm. wide; margins undulate. Peduncle terete, glabrous, 2–6 mm. in diameter, up to 70 cm. long, with deciduous leaves all along. Inflorescence a lax panicle, much exserted above the leaves; floral bracts lanceolate, glabrous. Pedicels articulated below to near the middle, sometimes near the base, 7–15 mm. long, 3–6(–10)-nate at the lower nodes. Perianth whitish, tinged or streaked white on the outside, urceolate; tepals reflexed distally, not ligulate, 6–10 mm. long, ± 2–3 mm. wide, 3-veined. Filaments scabrid to papillose, fusiform, widest in upper half, ± 5 mm. long; anthers versatile, 1.5–2 mm. long, twisted after anthesis. Ovary narrowing towards the base but not stipitate as in the previous species; style straight, ± 6 mm. long. Capsule triquetrous, 5–11 mm. long, emarginate. Seeds disc-shaped, 2–3 mm. in diameter. Fig. 2/3–5.

FIG. 2. *CHLOROPHYTUM ANDONGENSE* — **1**, habit, × $^{2}/_{9}$; **2**, fruiting node, × $^{2}/_{3}$. *CHLOROPHYTUM VIRIDESCENS* — **3**, habit, × $^{2}/_{9}$; **4**, flower, × 2; **5**, fruiting node, × $^{2}/_{3}$. *CHLOROPHYTUM NYASAE* — **6**, flower, × 2. 1, 2, from *Maitland* 1322; 3–5, from *Bidgood, Abdallah & Vollesen* 1993; 6, from *Richards* 17794. Drawn by Margaret Tebbs.

UGANDA. Acholi District: 6 km. S. of Naam Okora, 30 Nov. 1957, *Langdale-Brown* 2404!
KENYA. Trans-Nzoia District: slopes of Elgon, above Endebess, 30 Apr. 1961, *Polhill* 410!; Nanyuki District: Sweet Waters Ranch, 26 Dec. 1964, *Gillett* 16562!; Teita District: Tsavo East National Park, 5 Jan. 1967, *Greenway & Kanuri* 12966!
TANZANIA. Masai District: Longido Mt., 12 Dec.1959, *Verdcourt* 2533!; Mbulu District: Lake Manyara National Park, Endabash [Endebash], 20 Jan. 1971, *Vesey-FitzGerald* 6907!; Handeni District: 30 km. S. of Handeni, 10 Mar. 1953, *Drummond & Hemsley* 1478!
DISTR. **U** 1; **K** 3, 4, 6, 7; **T** 2, 3; Cameroon
HAB. Rocky outcrops, on shallow black clayish soils; 500–2700 m.

SYN. *C. macrocladum* Poelln. in Portug. Acta Biol., sér. B, 1: 307 (1946). Type: Tanzania, Handeni District, Useguha, Mzinga [Msinga] to Pongwe, *Peter* 7362 (B, holo.!)
[*C. andongense* sensu Polhill in Journ. E. Afr. Nat. Hist. Soc. 24: 12 (1962); Hanid in U.K.W.F.: 683 (1974), *non* Baker]

NOTE. See under *C. andongense*. Chromosome number: 2n=16, based on countings from Kenya, cf. Nordal et al. in Mitt. Inst. Allg. Bot. Hamburg 23 (1990).

3. **C. nyasae** (*Rendle*) *Kativu* in Nordic Journ. Bot. 13: 63 (1993), as '*nyassae*'. Type: Malawi, Mt. Mulanje, *Whyte* (BM, holo.!)

Plants 18–140(–200) cm. high. Rhizome vertical, short, covered with a dense mass of roots; roots, spongy, extensive, without tubers. Leaves glabrous, closely veined, with prominent midrib and margins, linear to linear-lanceolate, clasping below, 9–70 cm. long, 0.4–1(–2.5) cm. wide; margins glabrous, scabrid or ciliate, not undulate. Peduncle terete, with bract-like leaves all along, glabrous, up to 80 cm. long. Inflorescence a loose panicle, sometimes only with a few branches; rhachis terete, glabrous; floral bracts ovate, scabrid or shortly ciliate. Pedicels articulated below to near the middle, up to 12 mm. long in fruit, 2–4-nate at the lower nodes. Perianth white or sometimes yellowish white; tepals 7–11 mm. long, 3-veined. Stamens slightly exserted; filaments filiform, 6 mm. long. Ovary sessile; style filiform, equal to longer than the stamens. Capsule obovoid, triquetrous, ± 8 mm. long. Seeds disc-shaped, 2 mm. in diameter. Fig. 2/6.

TANZANIA. Buha District: Mkuti R., 42 km. from Kigoma on Kasulu road, 14 July 1960, *Verdcourt* 2818!; Njombe District: 24 km. N. of Sunji, 18 Jan. 1968, *Sturtz* 78! & Kipengere Mts., Mtorwi Peak, 13 Jan. 1957, *Richards* 7721!
DISTR. **T** 4, 7; Zambia and Malawi
HAB. In montane grassland and scrubland, often in gullies and around water courses, on stony ground; 800–2800 m.

SYN. *Anthericum nyasae* Rendle in Trans. Linn. Soc., Bot., ser. 2, 4: 52 (1894); P.O.A. C: 138 (1895); Baker in F.T.A. 7: 481 (1898)
Chlorophytum glabriflorum C.H. Wright in K.B. 1906: 170 (1906). Type: Malawi, Mt. Mulanje, Tuchila Plateau, *Purves* 17 (K, holo.!, BM, iso.!)

4. **C. ruahense** *Engl.* in E.J. 28: 361 (1900). Type: Tanzania, Iringa District, Uhehe, on a hanging cliff on Ruaha R., *Goetze* 461 (B, holo.!)

Plants 25–70 cm. high. Rhizome vertical, bearing fibrous remains of old leaf-bases; roots spongy, thick, without tubers. Rosette-leaves reduced to a few glabrous cataphylls. Peduncle terete, glabrous, with large lanceolate clasping glabrous leaves, up to 15 cm. long and ± 1 cm. wide all along. Inflorescence paniculate, lax; rhachis terete, glabrous; floral bracts narrowly lanceolate. Pedicels articulated above the middle, 5–15 mm. long in fruit, 2–4-nate at the lower nodes. Perianth greenish, ± 6.5 mm. long; tepals reflexed when dry, 3-veined. Stamens 5–6 mm. long; filaments terete, much longer than the anthers. Ovary sessile; style ± 6 mm. long. Capsule shallowly lobed, rounded in cross-section. Seeds ± irregularly folded, 1.5 mm. in diameter. Fig. 3.

FIG. 3. *CHLOROPHYTUM RUAHENSE* — **1**, base of plant, × $^1/_2$; **2**, upper part of flowering plant, × $^1/_2$; **3**, flower, x 5. All from *Strid* 2502. Drawn by Eleanor Catherine.

TANZANIA. Kilosa District: Usagara, SW. of Kidete, 5 Dec. 1895, *Peter* 32797!; Iringa District: Uhehe, on a hanging cliff on Ruaha R., 10 Jan. 1899, *Goetze* 461!
DISTR. **T** 6, 7; Zambia
HAB. In woodland, often by water courses, on grey laterite; ± 700 m.

SYN. *Anthericum peteri* Poelln. in F.R. 51: 138 (1942). Type: Tanzania, Kilosa District, Usagara, SW. of Kidete, *Peter* 32797 (B, holo.!)

NOTE. This species is very rare and in East Africa only known from the two type collections. Specimens from Zambia (illustrated in fig. 3) has shorter broader leaves (up to 5 cm. wide) but is otherwise very similar and is considered conspecific. More material is needed.

5. **C. stolzii** (*K. Krause*) *Kativu* in Nordic Journ. Bot. 13: 64 (1993). Type: Tanzania, Rungwe District, Kyimbila, Mulinda, *Stolz* 339 (B, holo., K, iso.!)

Plants robust, 85–250 cm. high. Rhizome moniliform, with corm-like elements covered by concentric leaf-scars; roots many, thin to slightly spongy, bearing long tubers at the tips. Leaves distichous or nearly so, firm, glabrous, sheathing at the base, broadly linear, up to 80 cm. long, 2–3.5 cm. wide; cataphylls papery. Peduncle terete, with bract-like leaves for its entire length, glabrous, up to 200 cm. long. Inflorescence unbranched, or with a few branches at the base, up to 70 cm. long; rhachis shallowly angled, glabrous; floral bracts ovate-lanceolate, glabrous, ± 5 mm. long. Pedicels articulated near or below the middle, up to 7 mm. long in fruit, 2–5-nate at the lower nodes. Perianth white with green keel, sometimes slightly pinkish, tepals 12–20 mm. long, 3–7 mm. wide, 3(–5)-veined. Stamens often arranged in groups, 3 + 3 or 2 + 4, somewhat shorter than the tepals; filaments filiform, 5–7 mm. long, shorter than the anthers. Style declinate, exserted. Capsule obovoid, nearly rounded in cross-section, shallowly transversely ridged, ± 10–12 mm. long, acute and not emarginate. Seeds irregularly folded, 1.5–2.5 mm. in diameter.

TANZANIA. Kigoma District: between Kigoma and Kalini, on road to Usumbura, 11 July 1960, *Verdcourt* 2797!; Ufipa District: 20 km. SE. of Sumbawanga, Molo Prison range and maize farm, 8 June 1980, *Hooper, Townsend & Mwasumbi* 1879!; Rungwe District: Tukuyu–Ipyana [Ipana], 22 Aug. 1933, *Greenway* 3603!
DISTR. **T** 4, 7; Zaire, Burundi, Angola, Zambia, Malawi and Mozambique
HAB. In open *Brachystegia* woodland and in tall grassland, often in wet places or along stream banks, on clayish or sandy soils; 800–2000 m.

SYN. *Acrospira asphodeloides* Baker in Trans. Linn. Soc., Bot., ser. 2, 1: 255, t. 34/4–7 (1878) & in F.T.A. 7: 477 (1898), *non Chlorophytum asphodeloides* C.H. Wright (1906). Lectotype, chosen by Marais & Reilly (1978): Angola, Pungo Andongo, *Welwitsch* 3777 (BM, lecto.!)
Debesia asphodeloides (Baker) Kuntze, Rev. Gen. Pl.: 708 (1891)
Albuca stolzii K. Krause in E.J. 57: 237 (1921)
Chlorophytum stolzii (K. Krause) Weim. in Bot. Notis. 1937: 434 (1937), *nom. invalid*, based on "*Ornithogalum stolzii*", an unpublished name
Anthericum russisiense Poelln. in F.R. 51: 70 (1942). Type: Burundi, Usumbura, *Keil* 278 (B, holo.!)
A. foliatum Poelln. in F.R. 53: 136 (1944). Type: Tanzania, Buha District, between Bagaga and Kasulu [Kassulo], *Peter* 37408 (B, holo.!, B, iso.!)
A. welwitschii Marais & Reilly in K.B. 32: 657 (1978). Based on *Acrospira asphodeloides*

NOTE. This taxon has been confused with another large and robust species, the Ethiopian/N. Kenyan *C. ducis-aprutii*. They differ, however, particularly in fruit and seed characters and are probably not closely related. *C. stolzii* is closely related to the pubescent *C. velutinum* Kativu, occurring close to the Tanzanian border in Malawi. Chromosome number: 2n=16 or 32, based on countings from Tanzania, cf. Nordal et al. in Mitt. Inst. Allg. Bot. Hamburg 23 (1990).

6. **C. nubicum** (*Baker*) *Kativu* in Nordic Journ. Bot. 13: 63 (1993); Nordal & Thulin in Nordic Journ. Bot. 13: 271 (1993); U.K.W.F., ed. 2: 317 (1994); Nordal in Fl. Ethiopia 6: 94 (1997). Type: Sudan, Jur, White Nile, Nyangara, *Petherick* (K, holo.!).

Plants loosely tufted, 10–50 cm. high. Rhizome thick, moniliform, often with fibrous remains of old leaf-bases; roots spongy and thick, sometimes reduced to sessile elongated tuber-like structures. Leaves hysteranthous, rosulate, linear, sheathing at the base, glabrous, much shorter than the peduncle, up to 25 cm. long and 5 mm. wide; margins minutely papillate, scabrid or ciliate. Peduncle terete, with small bract-like leaves along its entire length, breaking up on drying. Inflorescence usually an unbranched lax raceme, sometimes with 1–3 branches at the base; floral bracts small, scarious, lanceolate to ovate, aristate. Pedicels articulated below the middle, 10–14 mm. long in fruit, 2–4-nate at the lower nodes. Perianth white with dull pinkish to brownish stripes on the outside; tepals 3(–5)-veined, 7–15 mm. long. Stamens shorter than the perianth; filaments, 4–5 mm. long, glabrous, filiform; anthers up to ± the same length, often curved on drying. Style as long as the stamens. Capsule subglobose, trigonous, ± 5 mm. long and 4 mm. wide, shallowly transversely ridged. Seeds irregularly folded, 2 mm. in diameter.

UGANDA. Acholi District: 3.5 km. SE. of Naam Okora, 30 Nov. 1957, *Langdale-Brown* 2404!; Karamoja District: Moruita, Jan. 1963, *Tweedie* 2538!; Teso District: Serere, Feb. 1933, *Chandler* 1127!

KENYA. Uasin Gishu District: Kipkarren, Mar. 1932, *Brodhurst Hill* 692!; Naivasha, "Sterndale", 13 May 1943, *Andrews* in *Bally* 4471!

TANZANIA. Mpanda District, 1960, *Holland* 19!; Chunya District: Lupa road, *Menzies* 365!; Iringa District: Ruaha National Park, 2 km. NNW. of Msembe at track to Mbagi, 6 Aug. 1970, *A. Bjørnstad* 498!

DISTR. **U** 1, 3; **K** 3; **T** 4, 7; Guinea, Nigeria, Cameroon, Central African Republic, Sudan, Ethiopia, Zambia, Malawi and Mozambique

HAB. In dry open woodland and bushed grassland, on stony shallow soils; 800–2000 m.

SYN. *Anthericum nubicum* Baker in J.L.S. 15: 301 (1876) & in F.T.A. 7: 484 (1898); F.P.S. 3: 263 (1956); F.W.T.A., ed. 2, 3: 97 (1968); E.P.A.: 1531 (1971); Hanid in U.K.W.F.: 678 (1974)

Chlorophytum tinneae Baker in J.L.S. 15: 333 (1876) & in F.T.A. 7: 509 (1898). Type: Sudan, Bongo, Jur [Djur]; Kotschy & Peyr., Pl. Tinn., t. 23B (1876)

Anthericum fibrosum Hutch., F.W.T.A. 2: 342 (1936), *nom. invalid.* Based on N. Nigeria, *Dalziel* 230 & 261

A. senussiorum Poelln. in F.R. 53: 133 (1944). Type: Chad, Plains of Diyangoni, *Chevalier* 6734 (B, holo.!)

Chlorophytum pleurostachyum Chiov. in Webbia 8: 16, fig. 5 (1951); E.P.A.: 1535 (1971). Lectotype, chosen by Nordal & Thulin (1993): Ethiopia, Sidamo, Mega, *Corradi* 4667 (FT, lecto.!)

7. **C. ducis-aprutii** *Chiov.* in Nuov. Giorn. Bot. Ital., n.s. 36: 370 (1929); Nordal & Thulin in Nordic Journ. Bot. 13: 262, fig. 2, 3A, 11C (1993); Nordal in Fl. Ethiopia 6: 95, fig. 190.3 (1997). Type: Ethiopia, Bale, Tomono, *Basile* 277 (FT, holo.!)

Very robust plants 60–200 cm., from a thick moniliform rhizome, carrying the fibrous remains of old leaf-bases; roots spongy, extensive, without tubers. Leaves several, basal, distichous (might appear more rosulate in older stages), linear to narrowly lanceolate, 45–100 cm. long, 1.5–3 cm. wide, acute, sheathing below, margin often shortly ciliate, with a ± distinct midrib. Peduncle terete, glabrous, up to 1 m. long and with a diameter of ± 1 cm. at the base, with clasping bract-like leaves, 4–15 cm. long, along its entire length. Inflorescence usually a simple raceme, up to 65 cm. long, sometimes with a few basal branches (when grown in a green house, the branching may become more extensive); rhachis glabrous; floral bracts, ovate, cuspidate, 10–30 mm. long, sometimes ciliate along margin. Pedicels articulated near or below the middle, green below and whitish above the articulation, glabrous, 4–9 mm. long, 2–4-nate at the lower nodes. Tepals spreading, subequal, 3-veined, white with greenish stripe on the outside, 12–17 mm. long, 4–6 mm. wide, the outer slightly narrower than the inner. Stamens subequal, as long as the perianth; filaments fusiform, glabrous, 8–10 mm. long; anthers 5–8 mm., slightly curved apically at anthesis. Style declinate, exserted. Capsule deeply 3-lobed in cross-section, smooth, 9–14 mm. long, 7–9 mm. wide, with the perianth persistent at the base. Seeds thin, flat, black, 2.5–4 mm. in diameter.

KENYA. Northern Frontier Province: Moyale, 31 Oct. 1952, *Gillett* 14109! & 46 km. from Garissa on Hagadera road, 29 May 1977, *Gillett* 21225!

DISTR. **K** 1; Eritrea and Ethiopia

HAB. In grassy slopes or thicket, on ± loamy, dark brown to reddish soils; 1200–3000 m.

SYN. *Anthericum burgeri* Cufod. in E.P.A.: 1530 (1971) pro parte, quoad specimen *Burger* 1025
Chlorophytum sp. nov. sensu Nordal et al. in Mitt. Inst. Allg. Bot. Hamburg 23: 557 (1990)

NOTE. See note under *C. stolzii*. Chromosome number: 2n=32 (tetraploid), based on material from Ethiopia, cf. Nordal et al. in Mitt. Inst. Allg. Bot. Hamburg 23 (1990).

8. **C. longifolium** *Baker* in J.L.S. 15: 327 (1876) & in F.T.A. 7: 507 (1898); F.P.S. 3: 268 (1956); E.P.A.: 1535 (1971); Nordal & Thulin in Nordic Journ. Bot. 13: 269 (1993); Nordal in Fl. Ethiopia 6: 94, fig. 190.2 (1997). Type: Ethiopia, Tigray, Chiré, Plains of Beless, *Quartin-Dillon* (P, holo.!)

Plants 50–105 cm. Rhizome thick, horizontal, moniliform, with concentric rings of old leaf-attachments, up to 10 cm. long, bearing fibrous remains of old leaf-bases; roots long, spongy, often somewhat swollen towards the tips, but without distinct tubers. Leaves rosulate, linear to narrowly lanceolate, often half-folded (canaliculate), 17–50(–70) cm. long, and 1–2(–3) cm. wide; margins ± undulate, ciliate. Peduncle terete, with bract-like leaves along its entire length, glabrous below, densely papillate-pubescent above, 35–70 cm. long. Inflorescence a subspicate raceme, unbranched or with 1–3 basal branches; rhachis densely pubescent; floral bracts linear, ciliate, lower ones up to 2 cm. long. Pedicels articulated at apex, up to 1.5–5 mm. long, often pubescent, (2–)3–5-nate at the lower nodes. Perianth white, bell-shaped to urceolate, constricted just above the ovary; tepals 8–11 mm. long, 2.5–4 mm. wide, 3-veined, scabrid at the tips, densely glandular papillate on the inside at the constriction. Stamens exserted; filaments fusiform, widest in upper half, glabrous to papillate, 10–13 mm. long; anthers 2–3 mm. long. Style declinate, ± as long as the stamens. Capsule oblong, deeply 3-lobed (triquetrous), smooth, 6–10 mm. long, slightly narrower than long. Seeds flat, 3–4 mm. in diameter. Fig. 4/1–2.

TANZANIA. Shinyanga, *Koritschoner* 2295!; Kondoa District: between Kondoa and Mondo, 20 Jan. 1962, *Polhill & Paulo* 1240!; Iringa District: Mdonyo R., 12 km. from Msembe, 4 May 1970, *Greenway & Kanuri* 14450!

DISTR. **T** 1, 5–7; Sudan, Ethiopia, Angola, Zambia, Zimbabwe, Botswana and Namibia

HAB. In woodland, often on hill slopes, on shallow, rocky soils. 850–2000 m.

SYN. *Anthericum longifolium* A. Rich. in Tent. Fl. Abyss. 2: 333 (1850), *nom. illegit.*, *non* Jacq. (1786)
A. drimiopsis Baker in J.L.S. 15: 301 (1876). Type: Mozambique, between Tete and Lupata, *Kirk* (K, holo.!)
Dasystachys falcata Baker in Trans. Linn. Soc., Bot., ser. 2, 1: 256 (1878) & in F.T.A. 7: 512 (1898). Type: Angola, Huila, *Welwitsch* 3793 (BM, holo.!)
D. drimiopsis (Baker) Benth. in G.P. 3: 789 (1883); Baker in F.T.A. 7: 510 (1898)
Chlorophytum papillosum Rendle in J.L.S. 30: 422 (1895); P.O.A. C: 140 (1895). Type: Tanzania, between Uyui and the coast, 1886, *W.E. Taylor* (BM, holo.!)
Dasystachys papillosa (Rendle) Baker in F.T.A 7: 514 (1898)
Chlorophytum welwitschii Poelln. in Portug. Acta Biol., sér. B, 1: 220 (1945). Based on *Dasystachys falcata* Baker
C. drimiopsis (Baker) Poelln. in Portug. Acta Biol., sér. B, 1: 231 (1945)
C. pleiophyllum Poelln in Portug. Acta Biol., sér. B, 1: 361 (1946). Type: Tanzania, Tabora District, Ngulu, from Malongwe towards Nyahua, *Peter* 34518 (B, holo.!)
C. poricolum Poelln. in Portug. Acta Biol., sér. B, 1: 363 (1946). Type: Tanzania, Dodoma District, Uyansi, near Chaya, *Peter* 34006 (B, holo.!)

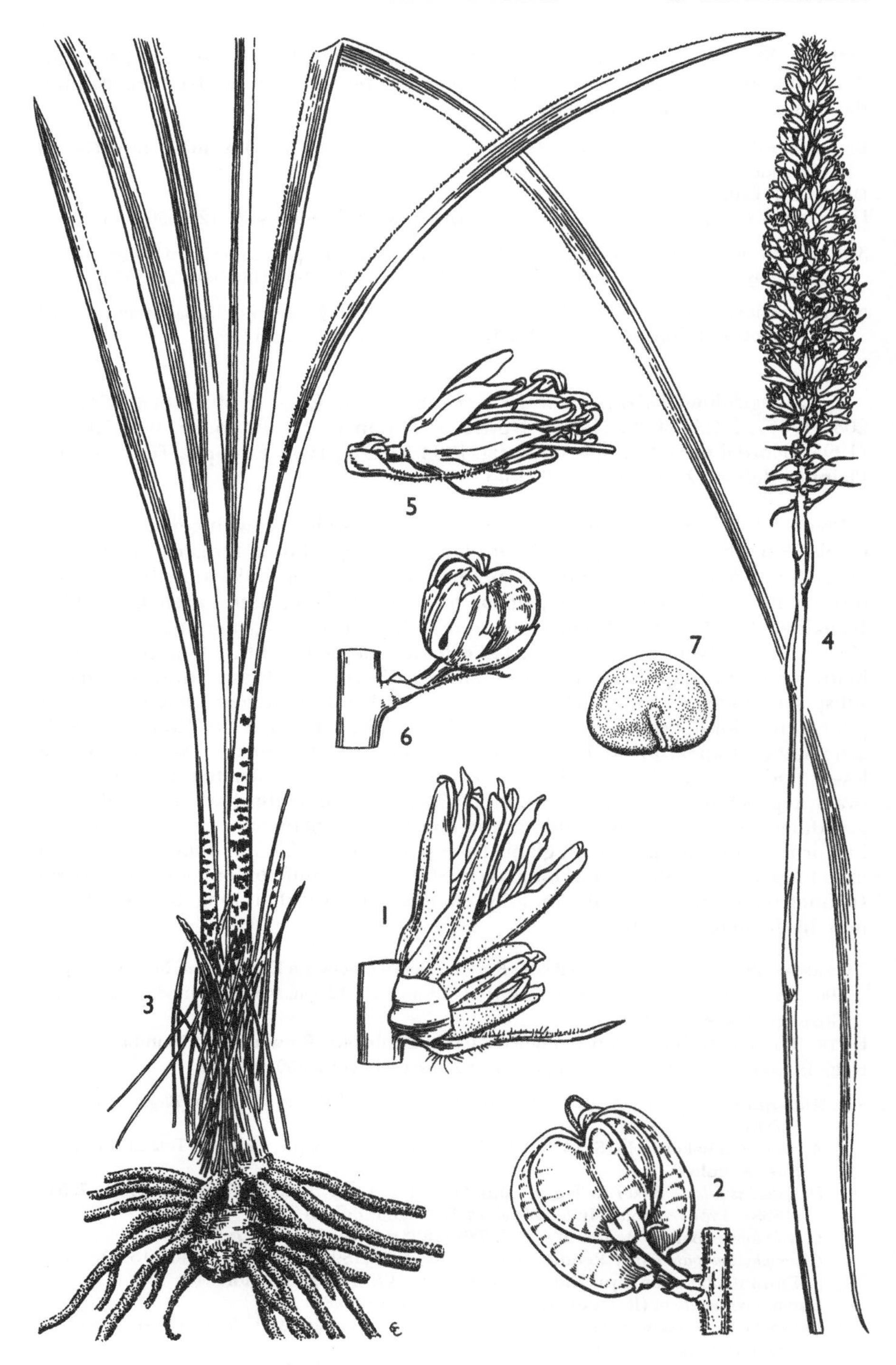

FIG. 4. *CHLOROPHYTUM LONGIFOLIUM* — **1**, flowering node, × 3; **2**, fruiting node, × 3. *CHLOROPHYTUM COLUBRINUM* — **3**, base of plant, × ²/₃; **4**, upper portion of peduncle with inflorescence, × ²/₃; **5**, flowering node, × 3; **6**, fruiting node, × 3; **7**, seed, × 6. 1, from *Pawek* 12316; 2, from *Philcox, Leppard & Dini* 8722; 3, from *Nuvunga* 656; 4, 5, from *White* 6638; 6, 7, from *Brummitt* 9317. Drawn by Eleanor Catherine.

NOTE. As *Anthericum longifolium* A. Rich. is illegitimate, the correct author citation is *C. longifolium* Schweinf. ex Baker, without referring to any basionym. The following citation often used should accordingly be avoided: "*C. longifolium* (A. Rich.) Schweinf. ex Baker". Chromosome number: 2n=16 based on material from Tanzania and Zimbabwe, cf. Nordal et al. in Mitt. Inst. Allg. Bot. Hamburg 23 (1990).

9. **C. colubrinum** (*Baker*) *Engl.* in Hochgebirgsfl. Trop. Afr. 1: 162 (1892) & in E.J. 28: 362 (1900). Type: Angola, Huila, *Welwitsch* 3784 (K, lecto.!, ?P, isolecto.)

Plants often in clumps, 20–150 cm. high. Rhizome thick, moniliform, horizontal, bearing fibrous remains of old leaf-bases; roots ± spongy, without tubers. Leaves subdistichous (distichous in young plants), sheathing at the base, linear to linear-lanceolate, firm, 12–75 cm. long, 0.5–2.5(–4) cm. wide, the margins scabrid to shortly ciliate; cataphylls and outer leaf-bases often reddish brown streaked or spotted reddish. Peduncle terete, glabrous, 20–95 cm. long, with bract-like leaves along its entire length. Inflorescence a narrow subspicate raceme, rarely with a few branches at the base, often dense, exserted above the leaves, 10–30 cm. long; floral bracts leafy, occasionally red spotted; rhachis angled, winged; floral bracts oblanceolate, aristate, purplish and ciliate on the margins, lower ones 8–20 mm. long. Pedicels articulated at the apex, glabrous, 1–5(–8) mm. long, solitary at the nodes. Perianth white, sometimes with greenish apices, ± bell-shaped, constricted above the ovary; tepals connate, cucullate, 6–10 mm. long, 1.5–3 mm. wide, 1-veined, papillate on the inside just above the ovary. Stamens exserted at anthesis; filaments filiform to slightly fusiform, longer than the tepals; anthers 1.5–2.5 mm., clasping the style at anthesis. Style straight to slightly declinate, slightly longer than the stamens. Capsule obovoid, triquetrous, emarginate, 4–8 mm. long, 3.5–5 mm. wide. Seeds disc-shaped, 2–3 mm. in diameter. Fig. 4/3–7.

TANZANIA. Dodoma District: Kazikazi, 14 Mar. 1933, *B.D. Burtt* 4620!; Mbeya District: Ruaha National Park, Magangwe ranger post, 8 Mar. 1972, *A. Bjørnstad* 1427!; Songea District: Songea airfield, 24 Feb. 1956, *Milne-Redhead & Taylor* 8879!

DISTR. **T** 1, 2, 4–8; Zaire, Rwanda, Burundi, Angola, Zambia, Malawi, Mozambique and Zimbabwe

HAB. In open woodland and wooded grassland, often on stony, gravelly, light soils, sometimes in seasonally flooded grassland; 500–3000 m.

SYN. *Dasystachys colubrina* Baker in Trans. Linn. Soc., Bot., ser. 2, 1: 256, t. 35/5–10 (1878) & in F.T.A. 7: 514 (1898)

D. campanulata Baker in Trans. Linn. Soc., Bot., ser. 2, 1: 256, t. 35/1–4 (1878) & in F.T.A. 7: 513 (1898). Type: Angola, Huila, *Welwitsch* 3783 (K, holo.!)

D. pleiostachya Baker in Trans. Linn. Soc., Bot., ser. 2, 1: 255 (1878) & in F.T.A. 7: 512 (1898). Type: Angola, Pungo Andongo, *Welwitsch* 3785 (BM, holo.!, P, iso.!)

Chlorophytum campanulatum (Baker) Engl. in Hochgebirgsfl. Trop. Afr. 1: 161 (1892)

C. pleiostachyum (Baker) T. Durand & Schinz in Consp. Fl. Afr. 5: 352 (1893)

Dasystachys crassifolia Baker in F.T.A. 7: 511 (1898). Types: Tanzania, between Lake Rukwa and Lake Tanganyika, *Nutt* (K, syn.) & Zambia, Urungu, Fwambo, *Nutt* (K, syn.)

D. decorata Baker in F.T.A. 7: 512 (1898). Type: Zambia, Fwambo, *Carson* 26 (K, holo.!)

D. grantii Benth. var. *engleri* Baker in F.T.A. 7: 513 (1898). Types: Tanzania, Tabora District, near Igonda, *Boehm* 18 & 162 (B, syn.!)

D. verdickii De Wild., Ann. Mus. Congo, Bot., sér. 4, 1: 10 (1902). Type: Zaire, Shaba, Lofoi, 1899, *Verdick* (BR, holo.!)

Chlorophytum dolichostachys Engl. & Gilg in Warb., Kunene-Sambesi Exped.: 188 (1903). Type: Angola, Longa, above Napalanka, *Baum* 611 (B, holo.!)

Dasystachys hockii De Wild. in B.J.B.B. 3: 264 (1911). Type: Zaire, Shaba, Feb. 1910, *Hock* (BR, holo.!)

Chlorophytum maculatum Dammer in E.J. 48: 365 (1912). Type: Tanzania, Kilwa District, Donde, near Umari-Kwa-Kinijalla, *Busse* 599 (B, holo., EA, iso.)

Dasystachys bequaertii De Wild. in F.R. 12: 294 (1913). Type: Zaire, Shaba, *Bequaert* 533 (BR, holo.!)

D. stenophylla R.E. Fr. in Wiss. Ergebn. Schwed. Rhod.-Kongo-Exped.: 226 (1916). Type: Zambia, Katakwe, *R.E. Fries* 1168 (UPS, holo.!)

Chlorophytum subpapillosum Poelln. in Portug. Acta Biol., sér. B, 1: 232 (1945). Type: Zaire, Lubembe valley, *Kassner* 2394 (B, holo.!, BR, iso.!)
C. crassifolium (Baker) Poelln. in Portug. Acta Biol., sér. B, 1: 355 (1946)
C. engleri (Baker) Poelln. in Portug. Acta Biol., sér. B, 1: 355 (1946), pro parte, excl. syn. *C. africanum* Engl.
C. engleri (Baker) Poelln. var. *angustifolium* Poelln. in Portug. Acta Biol., sér. B, 1: 356 (1946). Type: Tanzania, Kigoma District, W. Malagarasi, *Peter* 35974 (B, holo.!, BR, iso.!)
C. nguluense Poelln. in Portug. Acta Biol., sér. B, 1: 360 (1946). Type: Tanzania, Tabora District, Ngulu, E. of Goweko, *Peter* 34851 (B, holo.!)
C. rubromarginatum Poelln. in Portug. Acta Biol., sér. B, 1: 365 (1946). Type: cultivated in Dresden from Tanzania, Njombe District, Uhehe, Lupembe, *Schlieben* (B, holo.)
C. decoratum (Baker) Marais & Reilly in K.B. 32: 659 (1978)
C. stenophyllum (R.E. Fr.) Marais & Reilly in K.B. 32: 662 (1978)

NOTE. The species varies considerably in size, cataphyll/leaf-base coloration, and indumentum of leaf-margins. This variation was observed both at inter-population and intra-population levels. High altitude forms may be dwarfed with intense blackish coloration (e.g. *Kerfoot* 1633 from **T** 7, Mbeya peak). Chromosome number: 2n=16 based on countings from Tanzania, Zambia and Zimbabwe, cf. Nordal et al. in Mitt. Inst. Allg. Bot. Hamburg 23 (1990).

10. **C. silvaticum** *Dammer* in E.J. 48: 365 (1912); Nordal & Thulin in Nordic Journ. Bot. 13: 276 (1993); U.K.W.F., ed. 2: 317 (1994); Thulin, Fl. Somalia 4: 44 (1995); Nordal in Fl. Ethiopia 6: 95 (1997). Type: Tanzania, Kilwa District, Donde, near Kwa Mpanda, *Busse* 1310 (B, holo.!, EA, iso.!)

Plants often clumped, 10–25 cm. Rhizome short, moniliform, densely covered with fibrous remains of old leaf-bases; roots spongy, fusiform, occasionally reduced to sessile or subsessile tuberous structures. Cataphylls ± membranous, often whitish with green veins and crisped membranous margin. Leaves rosulate, linear-lanceolate, ± canaliculate, glabrous, 5–40 cm. long, (2–)5–15 mm. wide; margins often undulate. Peduncle without leaves, slender, terete, glabrous to slightly pubescent, particularly in upper part, up to 20 cm. long. Inflorescence a dense, subspicate raceme, very rarely forked at the base, 3–12 cm. long; floral bracts linear to lanceolate, usually greenish, rarely reddish, up to 10 mm. long, ciliate. Pedicels articulated at the apex, short, 1–3(–7) mm. long, solitary at the nodes. Perianth white, ± bell-shaped, slightly constricted above the ovary; tepals connate, cucullate, 5–7 mm. long, 1–2 mm. wide, 1-veined, papillose inside at the constriction. Stamens exserted; filaments filiform, 3–4 mm. long; anthers small, 1 mm. long, versatile, twisted after anthesis, much shorter than the filaments. Style exserted, ± straight. Capsule triquetrous, deeply lobed, 2–4 mm. long, 3–6 mm. wide. Seeds disc-shaped, ± 2.5 mm. in diameter.

UGANDA. Karamoja District: 32 km. N. of Kacheliba, 7 May 1953, *Padwa* 69! & Moroto, May 1956, *J. Wilson* 242! & Apr. 1959, *Phillip* 555!
KENYA. Northern Frontier Province: Ndoto Mts., Nguronit Mission Station, 31 Oct. 1978, *M.G. Gilbert & Gachathi* 5244!; W. Suk District: Kacheliba, 10 June 1959, *Symes* 611!; Masai District: 3 km. Athi R. to Namanga, 19 Nov. 1960, *Verdcourt* 3007!
TANZANIA. Musoma District: Serengeti, SW. of Klein's Camp, 23 Dec. 1969, *Greenway* 13893!; Mbulu District: Tarangire National Park, track near Park Warden's camp, 2 Dec. 1979, *Richards* 24881!; Iringa District: Ruaha National Park, Trekamboya track, 12 km. from Msembe, 12 Dec.1970, *Greenway & Kanuri* 14781!
DISTR. **U** 1; **K** 1–7; **T** 1–5, 7, 8; Ethiopia, Somalia, Zambia, Malawi, Mozambique and Zimbabwe
HAB. In open woodland (*Acacia-Commiphora* or *Brachystegia*) and grassland, often on ± degraded, badly drained, loamy or sandy, blackish or reddish soils; 100–2300 m.

SYN. *Dasystachys debilis* Baker in F.T.A. 7: 513 (1898); E.P.A.: 1537 (1971), *non Chlorophytum debile* Baker (1878). Type: Tanzania, Kilimanjaro, *Johnston* (K, holo.!)
D. gracilis Baker in F.T.A. 7: 510 (1898), *non Chlorophytum gracile* De Wild. (1911). Type: Kenya, Teita District, Mbuyuni, *Scott-Elliot* 6193 (BM, holo.!)
Chlorophytum papilliferum Poelln. in Portug. Acta. Biol., sér. B, 1: 231 (1945). Type: Malawi, Namadzi [Namasi], *Cameron* 80 (B, holo.!)

C. bakeri Poelln. in Portug. Acta. Biol., sér. B, 1: 351 (1946); Polhill in Journ. E. Afr. Nat. Hist. Soc. 24: 12 (1962); Hanid in U.K.W.F.: 683 (1974); Blundell, Wild Fl. E. Afr., reprint: 421, fig. 145 (1992). Based on *Dasystachys debilis* Baker
C. basitrichum Poelln. in Portug. Acta Biol., sér. B, 1: 352 (1946). Type: Tanzania, Dodoma District, Uyansi, near Chaya [Tschaya], *Peter* 34021 (B, holo.!)
C. comatum Poelln. in Portug. Acta. Biol., sér. B, 1: 354 (1946). Type: Tanzania, Tabora District, Ngulu, W. of Malongwe, *Peter* 34576 (B, holo.!, B, iso.!)
C. inexpectatum Poelln. in Portug. Acta. Biol., sér. B, 1: 357 (1946). Type: Tanzania, Dodoma District, Uyansi, Chaya, *Peter* 34027 (B, holo.!)
C. robustum Poelln. in Portug. Acta. Biol., sér. B, 1: 364 (1946). Type: Tanzania, Pare District, between Lembeni and Same, *Peter* 11450 (B, holo.!)
C. sp. near *bakeri* sensu Polhill in Journ. E. Afr. Nat. Hist. Soc. 24: 13, fig. 11 (1962)

NOTE. Chromosome number: 2n=16 based on countings from Kenya, cf. Nordal et al. in Mitt. Inst. Allg. Bot. Hamburg 23 (1990).

11. **C. africanum** (*Baker*) *Engl.* in E.J. 15: 470 (1892) & P.O.A. C: 140 (1895); Marais & Reilly in K.B. 32: 658 (1978). Lectotype, chosen by Benth., G.P. 3: 789 (1883): Tanzania, Tabora District, Rubuga [Rubugwa], *Grant* (K, lecto.!)

Plants often clumped, 25–50 cm. high. Rhizome horizontal, moniliform, bearing fibrous remains of old leaf-bases; roots extensive, spongy, without tubers. Leaves rosulate, linear-lanceolate to lanceolate, firmly membranous, glabrous, 20–50 cm. long, 1.5–3(–5) cm. wide; cataphylls and leaves with red-cilate margins and sometimes also veins. Peduncle terete, shorter than the leaves, glabrous to pubescent. Inflorescence a moderately dense, subspicate raceme, unbranched or with a single basal branch, 7–20 cm. long; rhachis subterete, angled, scabrid to shortly pubescent; floral bracts lanceolate, softly awned, red-ciliate on the margins, lower ones up to 15 mm. long. Pedicels articulated at the apex, up to 3 mm. long in fruit, solitary at the nodes. Perianth white, ± bell-shaped and slightly constricted above the ovary; tepals connate, 5–7 mm. long, narrow, 1-veined, cucullate, papillate on the inside just above the ovary. Stamens exserted; filaments fusiform, dilated in upper part, 4–6.5 mm. long; anthers versatile, 1.5–3 mm. long, twisted after anthesis. Style exserted, declinate. Capsule triquetrous, deeply lobed, emarginate, 2–4 mm. long, 3–5.5 mm. wide. Seeds saucer-shaped, 3 mm. in diameter.

TANZANIA. Mpanda, 29 Nov. 1949, *Shabani* 61!; Dodoma District: Chenene, 23 Jan. 1962, *Polhill & Paulo* 1240!; Iringa District: Ruaha National Park, Magangwe Ranger Post, 13 Dec. 1972, *A. Bjørnstad* 2054!
DISTR. **T** 1, 4, 5, 7; Zambia
HAB. In *Acacia-Commiphora* woodland, rocky hillsides, on red sandy soils; 1300–1400 m.

SYN. *Caesia africana* Baker in Trans. Linn. Soc. 29: 160, t. 103A (1875)
Dasystachys grantii Benth., G.P. 3: 789 (1883); Baker in F.T.A. 7: 513 (1898), *nom. illegit.* Type as above
Chlorophytum marginatum Rendle in J.L.S. 30: 422 (1895); P.O.A. C: 140 (1895). Type: Tanzania, Tabora District, Uyui, 3 Feb. 1887, *W.E. Taylor* (BM, holo.!)
Dasystachys marginata (Rendle) Baker in F.T.A. 7: 512 (1898)
D. africana (Baker) T. & H. Durand, Syll. Fl. Cong.: 569 (1909)
Chlorophytum marginatum Rendle var. *hispidulum* Poelln. in Portug. Acta Biol., sér. B, 1: 360 (1946). Type: Tanzania, Tabora District, Ngulu, between Malongwe and Tura, *Peter* 34711 pro parte (B, holo.!)
C. setosum Poelln. in Portug. Acta. Biol., sér. B, 1: 366 (1946). Type: Tanzania, Musoma District, E. of Ikoma, Ermessa, *Jaeger* 350 (B, holo.!)
C. turuense Poelln. in Portug. Acta Biol., sér. B, 1: 367 (1946). Type: Tanzania, Dodoma District, Turu, E. of Itigi, towards Bangayega, *Peter* 33717 (B, holo.!)
C. turuense Poelln. var. *micranthum* Poelln. in Portug. Acta Biol., sér. B, 1: 368 (1946). Type: Tanzania, Dodoma District, Kilimatinde, *Prittwitz* 34 (B, holo.!)

NOTE. This species is closely related to *C. silvaticum*, and may only represent a ciliate-leaved form of that species. *C. macrorrhizum* is another ciliate-leaved and closely connected species. More material is needed to settle the taxonomic delimitation in this group. Chromosome number: 2n=80, based on countings from Kenya and Tanzania, cf. Nordal et al. in Mitt. Inst. Allg. Bot. Hamburg 23 (1990). The taxon appears to be a decaploid derivative within the *C. silvaticum* group.

12. **C. macrorrhizum** *Poelln.* in Portug. Acta Biol., sér. B, 1: 358 (1946). Type: Tanzania, Dodoma District, Turu, E. of Itigi, towards Bangayega, *Peter* 33856 (B, holo.!)

Plants small, up to 15 cm. high. Rhizome horizontal, moniliform, covered by a mass of dense roots, bearing some fibrous remains of old leaf-bases; roots thick, short, spongy, occasionally reduced to subsessile tubers. Leaves ± prostrate, lanceolate, obtuse to acute, densely pubescent on nerves beneath, glabrous above, up to 10 cm. long, 1–2 cm. wide, margins occasionally wavy, ciliate; cataphylls prominent, reddish. Peduncle short, terete, purplish pubescent. Inflorescence dense, unbranched, up to 3.5 cm. long; floral bracts lanceolate, purple-ciliate, 5–10 mm. long. Pedicels articulated at the apex, up to 1.5 mm. long in fruit, solitary at the nodes. Perianth campanulate; tepals connate, white, papillose at the constriction, up to 5 mm. long, 1-veined. Stamens as long as the perianth; filaments filiform, up to 5 mm. long; anthers 1 mm. long. Style straight, exserted. Capsule small, triquetrous, ± 1.5 mm. long, 3 mm. wide. Seeds few, flat to saucer-shaped, ± 1.5 mm. in diameter.

TANZANIA. Ufipa District: new Sumbawanga–Mbala [Abercorn] road, 32 km. from Mbala [Abercorn], 25 Nov. 1960, *Richards* 13627! & Mchata [Machata] Mts., 13 Nov. 1958, *Richards* 10298!; Dodoma District: Turu, E. of Itigi, towards Bangayega, 1 Jan. 1926, *Peter* 33856!

DISTR. **T** 4, 5; Angola and Zambia

HAB. In swampy and seasonally very wet areas, on lateritic and clay soils; 1250–1500 m.

NOTE. The taxonomic delimitation is not clear. See under *C. africanum*, above.

13. **C. leptoneurum** (*C.H. Wright*) *Poelln.* in Portug. Acta Biol., sér. B, 1: 358 (1946). Type: Malawi, Nyika Plateau, Nacheri, *McClounie* 89 (K, holo.!, B, iso.!)

Plants 5–8(–12) cm. high, growing in clumps. Rhizome horizontal, long and narrow, stoloniferous, with subterranean runners producing new plants; roots reduced to sessile elongated tubers. Leaves rosulate, ± prostrate, linear to oblanceolate, membranous, pubescent on the adaxial surface, glabrous on the abaxial surface, 3–10 cm. long, 0.5–1.5(–2) cm. wide; margins ciliate. Peduncle terete, closely ribbed, pubescent, 1.3–2.2 cm. long. Inflorescence dense, unbranched; floral bracts lanceolate, ciliate. Pedicels articulated at the apex, up to 3 mm. long, solitary at the nodes. Perianth slightly campanulate; tepals connate, white, 1-veined. Stamens exserted; filaments filiform, longer than the anthers. Ovary sessile to subsessile; style as long as the stamens. Capsule subglobose, 2.5 mm. long, 3 mm. wide. Seeds saucer-shaped, ± 1.5 mm. in diameter.

TANZANIA. Ufipa District: Kalambo Falls, 11 Feb. 1965, *Richards* 19647!; Mpwapwa District: Ugogo, Gulwe, 8 Dec. 1925, *Peter* 32927!; Songea District: 9 km. W. of Songea, 4 Jan. 1956, *Milne-Redhead & Taylor* 8133!

DISTR. **T** 4, 5, 8; Zaire, Angola, Zambia, Malawi and Zimbabwe

HAB. In open *Brachystegia* woodland and grassland, on bare patches, often on sandy soils; 800–2600 m.

SYN. *Dasystachys leptoneura* C.H. Wright in K.B. 1908: 440 (1908)

Chlorophytum tenellum Poelln. in Portug. Acta Biol., sér. B, 1: 339 (1946). Type: Tanzania, Mpwapwa District, Ugogo, Gulwe, *Peter* 32865 (B, holo.!)

Verdickia katangensis De Wild., Ann. Mus. Congo, Bot., sér. 4, 1: 7 (1902). Type: Zaire, Shaba, Lukafu, *Verdick* 329 (BR, holo.!)

14. **C. collinum** (*Poelln.*) *Nordal* in Mitt. Inst. Allg. Bot. Hamburg 23: 539 (1990). Type: Tanzania, Morogoro District, Uluguru Mts., *Schlieben* 3183 (B, holo.!)

Plants 10–30(–40) cm. high. Rhizome short, vertically moniliform, densely covered with fibrous remains of old leaf-bases; roots spongy and fusiform. Cataphylls and leaves sheathing; leaves rosulate, linear-lanceolate, rarely broadly lanceolate, glabrous, up to 15(–30) cm. long, 5–10(–25) mm. wide; margins often undulate. Peduncle without leaves, slender, terete, 5–20(–30) cm. long. Inflorescence a simple raceme, elongated to ± capitate; rhachis pubescent; floral bracts lanceolate, 2–5 mm. long. Pedicels articulated at the apex, pubescent, 4–6 mm. long, solitary at the nodes, in robust specimens sometimes 2-flowered. Perianth white, greenish keeled on the outside, slightly urceolate to star-shaped; tepals (4–)6–9 mm. long, 2–2.5 mm. wide, 1-veined. Stamens slightly shorter than the tepals; filaments filiform, 4–6 mm. long; anthers 2–3 mm. long. Style declinate. Capsule triquetrous, emarginate, 3–5 mm. long, 4–5.5 mm. wide. Seeds disc-shaped, ± 2.5 mm. in diameter.

TANZANIA. Morogoro District: 5 km. S. of Mziha, 21 Nov. 1955, *Milne-Redhead & Taylor* 7097!; Mbeya District: Kitulo [Elton] Plateau, 26 Jan. 1961, *Richards* 14193!; Lindi District: 9.5 km. S. of Mbemkuru R., *Milne-Redhead & Taylor* 7475!

DISTR. **T** 3, 5–8; not known elsewhere

HAB. Montane grassland, deciduous bushland, riverine thicket; 100–3000 m.

SYN. *Anthericum collinum* Poelln. in F.R. 51: 119 (1942)

A. pseudopapillosum Poelln. in F.R. 51: 31 (1942). Type: Tanzania, Morogoro District, Useguha, between Ngotsche and Ngerengere R., *Peter* 7177 (B, holo.!)

Chlorophytum angustiracemosum Poelln. in Portug. Acta Biol., sér. B, 1: 278 (1946). Type: Tanzania, Handeni District, Useguha, Mlembule to Mzinga [Msinga], *Peter* 7326 (B, lecto.!, mixed collection with *C. africanum*)

C. pachyrrhizum Poelln. in Portug. Acta Biol., sér. B, 1: 319 (1946). Type: Tanzania, Mpwapwa District, Ugogo, Gulwe, *Peter* 32867 (B, holo.!)

NOTE. This species appears to link the bell-flowered, formerly *Dasystachys*, species with the ± open flowered *Chlorophytum* species. Except for the flower shape it is very close to *C. silvaticum*. The capitate form seems to occur only at high altitudes. A form with broad ciliate prostrate leaves, which might deserve taxonomic recognition, is known from **T** 7, Iringa District (*Greenway & Kanuri* 14777). Chromosome number: 2n=16, based on countings from Tanzania, cf. Nordal et al. in Mitt. Inst. Allg. Bot. Hamburg 23 (1990).

15. **C. affine** *Baker* in Trans. Linn. Soc. 29: 160, t. 104 (1875) & in J.L.S. 15: 327 (1876); P.O.A. C: 140 (1895); Baker in F.T.A. 7: 507 (1898); Hanid in K.B. 29: 588 (1974); U.K.W.F., ed. 2: 318, t. 146 (1994). Type: Tanzania, Tabora District, Unyamwezi, Rubuga, *Grant* (K, holo.!)

Plants variable, slender to moderately robust, often in dense clumps, 10–40 cm. high. Rhizome horizontal, extensive, moniliform, occasionally branched, carrying the fibrous remains of old leaf-bases; roots thin and wiry, with whitish subglobose tubers, 1–3 cm. long. Leaves distichous, erect to falcate, often canaliculate, sometimes finely undulate, linear to narrowly lanceolate, firm, 10–40 cm. long, 4–20 mm. wide, margins and often midribs as well ciliate; leaf-bases and cataphylls blotched with greenish brown to brownish red spots or bands. Peduncles upright or geniculately curved (arcuate) at the base, angled, glabrous to shortly hairy below, densely papillate to pubescent above. Inflorescence a short, 3–13 cm. long, simple raceme; rhachis angled, pubescent, sometimes sinuate, scabrid to pubescent; floral bracts lanceolate, cuspidate, ciliate on the margins, lowermost up to 7 mm. long. Pedicels suberect, articulated near the middle or in the lower half to near base, ± papillate, 3.5–10 mm. long in fruit, 1–3-nate at the lower nodes. Flowers white, star-shaped; tepals patent, (6–)8–11 mm. long, 2–4 mm. wide, 3(–5)-veined, with a greenish brown stripe on the outside. Stamens equal to shorter than the tepals; filaments filiform to fusiform, glabrous, 3–6 mm. long; anthers 2–5 mm. long, twisted after anthesis. Style straight, slightly longer than the stamens. Capsule deeply triquetrous, smooth to slightly transversely ridged, 4–5 mm. long, 5–7 mm. wide, emarginate. Seeds saucer-shaped to flat, 2–3 mm. in diameter.

var. **affine**; Hanid in K.B. 29: 588, fig. 1 (1974)

Plants robust. Leaves 11–22 mm. wide, rarely narrower. Peduncle straight.

UGANDA. Karamoja District: Lodoketemit [Lodoketeminit], 20 May 1963, *Kerfoot* 4945! & Moroto, 5 Oct. 1952, *Verdcourt* 758!

KENYA. W. Suk District: Kacheliba, 10 June 1959, *Symes* 609! & Kongelai [Kongoli] road, July 1961, *Lucas* 187!; Machakos District: 46 km. on Thika–Kagondi [Kangonde] road, Kithimani, 29 Apr. 1967, *Hanid & Kiniaruh* 399!

TANZANIA. Biharamulo District: Ruiga R. Forest Reserve, 5 Dec. 1956, *Gane* 93!; Kondoa District: 19 km. S. of Kondoa, 18 Jan. 1962, *Polhill & Paulo* 1212!; Songea District: by Mkurira R., 26 Dec. 1955, *Milne-Redhead & Taylor* 7904!

DISTR. **U** 1; **K** 2, 4; **T** 1, 2, 4, 5, 7, 8; Senegal, Mali, Guinea, Ghana, Benin, Nigeria, Cameroon, Central African Republic, Zaire, Zambia and Malawi

HAB. In ± open woodland, wooded grassland or thornscrub, often overgrazed and eroded sites with bare patches, usually on sandy soils, more rarely on black cotton soils and hard pans; 900–2800 m.

SYN. *Anthericum pubirachis* Baker in J.L.S. 15: 302 (1876) & in F.T.A. 7: 481 (1898); F.W.T.A., ed. 2, 3: 96 (1968), as '*pubirhachis*'; Hanid in U.K.W.F. 678 (1974). Type: N. Nigeria, Nupe, *Barter* (K, holo.!)

A. taylorianum Rendle in J.L.S. 30: 415 (1895); P.O.A. C: 139 (1895); Baker in F.T.A. 7: 492 (1898). Type: Tanzania, Tabora District, Unyamwezi, Uyui, Feb. 1887, *W.E. Taylor* (BM, holo.!)

A. turuense Poelln. in F.R. 51: 134 (1942). Type: Tanzania, Dodoma District, Turu, between Itigi and Bangayega, *Peter* 33725 (B, holo.!)

Chlorophytum conspicuum Poelln. in Portug. Acta Biol., sér. B, 1: 286 (1946). Type: Tanzania, Shinyanga, *B.D. Burtt* 5078 (B, holo.!, BR, K, iso.!)

C. pubescens Poelln. in Portug. Acta Biol., sér. B, 1: 328 (1946). Type: Tanzania, Tabora District, Unyanyembe, Tabora to Kwihala, *Peter* 35336 (B, holo.!)

C. sp. B sensu Hanid in U.K.W.F.: 685 (1974)

[*C. uvinsense* sensu U.K.W.F., ed. 2: 318 (1994), *non* Poelln.]

var. **curviscapum** (*Poelln.*) *Hanid* in K.B. 29: 588, fig. 2 (1974). Type: Tanzania, Dodoma District, Uyansi, near Chaya, *Peter* 45827 (B, holo.!)

Plants usually small and slender. Leaves up to 9 mm. wide, rarely wider. Peduncle arcuate near the base, often prostrate. Fig. 5.

UGANDA. Acholi District: Adilang, Apr. 1943, *Purseglove* 1538!

KENYA. Northern Frontier Province: Lolokwi [Lolokwe], Ol Donyo Sabachi valley, *M.G. Gilbert* 5352!; Trans-Nzoia District: Elgon, Endebess, 12 May 1955, *Rayner* 611!; Nairobi, Uhuru Highway and Langata Road, 8 May 1975, *Kabuye & Ng'weno* 503!

TANZANIA. Dodoma District: 19 km. from Itigi on Manyoni road, 25 Apr. 1962, *Polhill & Paulo* 2172!; Mbeya District: Chimala, 17 Feb. 1979, *Leedal* 5387!; Songea District: Songea airfield, 20 Jan. 1956, *Milne-Redhead & Taylor* 8290!

DISTR. **U** 1; **K** 1–4, 6; **T** 1, 2, 4–8; Ghana, Cameroon, Chad, Sudan, Ethiopia, Somalia, Zambia, Malawi and Zimbabwe

HAB. In open woodland, bushland, wooded grassland, or on rocky outcrops, often on disturbed and eroded ground, on shallow soils often overlying limestone, sometimes seasonally wet, on sands, stony soils and heavy black cotton clays; 950–2150 m.

SYN. *Anthericum ledermannii* Engl. & K. Krause in E.J. 45: 127 (1910). Types: Cameroon, Garoua, *Ledermann* 4476 (B, syn.!) & Garoua, Schuari, *Ledermann* 5009 (B, syn.!)

A. pendulum Engl. & K. Krause in E.J. 45: 128 (1910). Type: Cameroon, Ngesik, *Ledermann* 4280a (B, holo.!)

A. deflexum A. Chev., Etudes Fl. Afr. Centr. Fr. 1: 315 (1913), *nom. nud.*

Chlorophytum tordense Chiov., Result. Sci. Miss. Stef.-Paoli, Coll. Bot.: 173 (1916); Polhill in Journ. E. Afr. Nat. Hist. Soc. 24: 14, fig. 14 (1962); E.P.A.: 1537 (1971); Hanid in U.K.W.F.: 684 (1974); Nordal & Thulin in Nordic Journ. Bot. 13: 277 (1993); Thulin, Fl. Somalia 4: 49 (1995); Nordal in Fl. Ethiopia 6: 102 (1997). Type: Somalia, Torda, Paoli 321 (FT, holo.!)

FIG. 5. *CHLOROPHYTUM AFFINE* var. *CURVISCAPUM* — **1**, habit, × $^2/_3$; **2**, flowering node, × 2; **3**, capsule, × 3; **4**, seed, × 4. Reproduced from K.B. 29, fig. 2 (1974). Drawn by M.A. Hanid.

Anthericum curviscapum Poelln. in F.R. 51: 122 (1942)
A. pubiflorum Poelln. in F.R. 51: 24 (1942). Type: Tanzania, Dodoma District, Turu, between Itigi and Bangayega, *Peter* 33754 (B, holo.!)
A. ciliare Poelln. in F.R. 51: 132 (1942). Type: Tanzania, Dodoma District, Uyansi, between Chaya and Turu, *Peter* 34195 (B, holo.!)

NOTE. Whether the two varieties deserve taxonomic rank is a question. Sometimes they are found together (e.g. in *Napper* 2119! from Kenya, W. Suk) and intermediates are found from time to time (particularly in the northern part of the distribution area). Drier habitats seem to favour the *curviscapum* variety. Up to now *C. tordense* has been regarded as a separate species adapted to the drier northern areas. In every character, however, it falls within the variation range of *C. affine* s.l. Chromosome number: 2n=16 based on countings from Kenya, cf. Nordal et al. in Mitt. Inst. Allg. Bot. Hamburg 23 (1990).

16. **C. pendulum** *Nordal & Thulin* in Nordic Journ. Bot. 13: 273, fig. 7, 11G (1993); Nordal in Fl. Ethiopia 6: 100, fig. 190.6 (1997). Type: Ethiopia, Bale, *Nordal, Melaku & Petros* 2294 (O, holo.!, ETH, iso.!)

Slender tufted plants, 15–40 cm. tall, from a short, sometimes moniliform rhizome, carrying the fibrous remains of old leaf-bases; roots thin and wiry, with tubers 1–2 cm. long. Leaves subdistichous, green to olive-green, linear, 20–45 cm. long, 4–9 mm. wide, sheathing below, erect but drooping in upper parts, somewhat canaliculate with a distinct midrib, margin distinctly ciliate; cataphylls ± membranous, sometimes with a characteristic white/green striping. Peduncle slender, lax, glabrous, leafless, 5–30 cm. long. Inflorescence a simple raceme; rhachis slightly papillose, often drooping; floral bracts, small, membranous, ovate, cuspidate, 1–4 mm. long, sometimes ciliate along margin. Pedicels articulated near the base, very thin, glabrous, 5–12 mm. long at anthesis, 2–4-nate at the lower nodes. Tepals patent, white, with greenish to brownish stripes on the outside, 5–9 mm. long, 3-veined. Stamens shorter than the perianth; filaments linear, scabrous to papillate, ± 5 mm. long; anthers 2 mm., curved at anthesis. Style straight, as long as the perianth. Capsule pendent, deeply 3-lobed, slightly transversely ridged, 10–15 mm. long, 6–10 mm. wide, triangular in longitudinal section, broadest at the distal end, emarginate, with remnants of the perianth at the base. Seeds thin, flat, black, 3–4 mm. in diameter.

KENYA. Northern Frontier Province: Moyale, 16 Apr. 1952, *Gillett* 12819! & 29 Sept. 1952, *Gillett* 13968!; Meru District: 19 km. ENE. of Isiolo on road to Mado Gashi, 16 May 1979, *Gillett* 19672!
DISTR. **K** 1, 4; Ethiopia
HAB. In woodland dominated by *Acacia, Combretum,* and/or *Commiphora,* in ± shade, on dark or red stony soils, sometimes on limestone; 1050–1200 m.

17. **C. fischeri** (*Baker*) *Baker* in F.T.A. 7: 506 (1898). Type: Tanzania, Kwimba District, between Magu and Kagehi, *Fischer* 591 (K, holo.!, B, iso.!)

Plants small, tufted, up to 20 cm. high. Rhizome short, moniliform, horizontal, bearing fibrous remains of old leaf-bases; roots fibrous, bearing distinct tubers, sometimes close to the rhizome, with patterned surface towards the tips. Leaves rosulate, few, linear-lanceolate, clasping below, minutely pubescent to papillate, particularly on the veins, 6–12 cm. long, 4–6 mm. wide, the bases and cataphylls with papillate to ciliate margins; cataphylls distally patent, base whitish to orange with green veins. Peduncle stiffly erect, never arcuate, terete, sometimes glabrous below, papillate to pubescent above, 7–10 cm. long. Inflorescence simple, open, exserted above the leaves, 6–20 cm. long; rhachis pubescent. Pedicels articulated near the middle, short, up to ± 4 mm. long in fruit, usually solitary at the nodes. Perianth white, papillose; tepals 3–5(–8) mm. long, 2 mm. wide, 3-veined. Stamens as long as the perianth; filaments filiform, scabrid, ± 3 mm. long, longer than the 1 mm. long versatile anthers. Style as long as the stamens. Capsule triquetrous, emarginate, ± 3 mm. long, 4–5 mm. wide. Seeds saucer-shaped, 1.5–2 mm. in diameter.

TANZANIA. Mpanda District: Mlala Hills, 27 Oct. 1959, *Richards* 11592!; Dodoma District: Manyoni, 4 Dec. 1931, *B.D. Burtt* 3527!; Morogoro District: 21 km. SSW. of Morogoro, 31 Dec. 1970, *Wingfield* 2281!; Mbeya District: Magangwe, 10 Dec. 1970, *Greenway & Kanuri* 14735!
DISTR. **T** 1, 4–8; Zambia and Malawi
HAB. Woodland and wooded grassland, on brown sandy loam, often on termite mounds; 450–2000 m.

SYN. *Anthericum fischeri* Baker in E.J. 15: 468 (1892); P.O.A. C: 139 (1895)
Chlorophytum montanum Poelln. in Portug. Acta Biol., sér. B, 1: 316 (1946). Type: Tanzania, Morogoro District, Uluguru Mts., on the southern hill, *Schlieben* 3086 (B, holo.!, BR, EA, K, iso.!)

NOTE. This taxon is closely related to *C. affine*. One specimen (*Bidgood, Abdallah & Vollesen* 1953 from **T** 8, Masasi District, 30 km. NW. of Masasi, Chiwale) is probably an undescribed taxon related to *C. fischeri*. It is 5–8 cm. tall, slender with ± prostrate leaves and a slightly winged and ciliate peduncle. More material is needed.

18. **C. bifolium** *Dammer* in E.J. 38: 66 (1905); E.P.A.: 1533 (1971); Nordal & Thulin in Nordic Journ. Bot. 13: 262 (1993); Thulin, Fl. Somalia 4: 45 (1995); Nordal in Fl. Ethiopia 6: 95 (1997). Type: Ethiopia, Bale, between Marta and Djaro, *Ellenbeck* 2042 (B, holo.!)

Small plants up to 15 cm. high, from a very short rhizome; roots reduced to a fascicle of elongated tubers, 1.5–2 cm. long. Leaves few, rosulate, lanceolate, 8–14 cm. long, ± 1.5 cm. wide, with a hyaline margin. Peduncle slender, scabrid, leafless, 5–7 cm. long. Inflorescence a simple raceme; rhachis ± pubescent; floral bracts narrow, membranous, up to 3 mm. long. Pedicels articulated near or slightly above the middle, suberect, glabrous, 3 mm. long at anthesis, usually solitary at the nodes. Flowers inconspicuous; tepals semipatent, whitish, 6 mm. long, 3-veined. Stamens shorter than the perianth. Capsule trigonous, indistinctly transversely ridged, 6–9 mm. long and 9–12 mm. wide, emarginate, with withered remnants of the perianth at the base. Seeds flat, slightly folded, 2 mm. in diameter.

KENYA. Northern Frontier Province: 2 km. N. of El Wak, 30 Apr. 1978, *M.G. Gilbert & Thulin* 1262!
DISTR. **K** 1; Ethiopia and Somalia
HAB. In *Acacia-Commiphora* bushland, with shallow, poorly drained soils, on limestone; 400–500 m.

SYN. [*Anthericum verruciferum* sensu Chiov., Result. Sci. Miss. Stef.-Paoli, Coll. Bot.: 174 (1916), pro parte quoad *Paoli* 942! & 1254!, see under *Anthericum jamesii*]

19. **C. somaliense** *Baker* in E.J. 15: 469 (1892); Nordal & Thulin in Nordic Journ. Bot. 13: 276 (1993); U.K.W.F., ed. 2: 318 (1994); Thulin, Fl. Somalia 4: 44 (1995); Nordal in Fl. Ethiopia 6: 95, fig. 190.4 (1997). Type: Somalia, near Mait, Mt. Ahl, *Hildebrandt* 1468 (B, holo.!)

Plants 30–70 cm. high. Rhizome short, moniliform, bearing fibres from previous leaves; roots thick and spongy throughout, or narrow near the base and enlarging into very long, conspicuous tubers. Leaves rosulate, erect or sometimes falcate with clasping bases, canaliculate, glabrous, often bluish to greyish green, narrowly lanceolate to linear, 20–40 cm. long, 1–3 cm. wide, margin undulate, cataphylls with a hyaline margin. Peduncle erect, leafless, 15–40 cm. Inflorescence a simple raceme; rhachis glabrous; floral bracts short, indistinct. Pedicels articulated in upper half, glabrous, ± 1 cm. at anthesis, elongating with age, solitary at the nodes. Flowers conspicuous, zygomorphic; tepals white with pale green midrib, 15–20 mm. long, 2–3 mm. wide, 3-veined, constricted and thus forming an urceolate structure around the ovary, outer parts of tepals reflexed at anthesis. Stamens longer than the perianth, exserted; filaments declinate, up to 15 mm.; anthers 2–4 mm., twisted after anthesis. Style declinate, exserted, as long as the stamens. Capsule with remnants of the perianth at the base, triquetrous, emarginate, variable in size, but often very large, (7–)10–20 mm. long, usually distinctly longer than wide. Seeds disc-shaped, 3–5 mm. in diameter. Fig. 6.

KENYA. Northern Frontier Province: Mandera, 5–10 km. SSE. of Ramu, 2 May 1978, *M.G. Gilbert & Thulin* 1327!; Turkana District: 64 km. SW. of Lodwar, Lorengipe, 11 Apr. 1954, *Hemming* 270!; Teita District: Irima, 13 Dec. 1966, *Greenway & Kanuri* 12736!

TANZANIA. Lushoto District: 5 km. NW. of Mombo, 29 Apr. 1953, *Drummond & Hemsley* 2280!

DISTR. **K** 1–4, 6, 7; **T** 3; Ethiopia and Somalia

HAB. In ± degraded *Acacia-Commiphora* bushland, also on treeless grassland; on rather shallow black cotton soil or red lateritic soils, but also in alluvial sands, often degraded and eroded, on granite, limestone, lava or gypseous rocks; 100–1700 m.

SYN. *C. tenuifolium* Baker in K.B. 1895: 228 (1895) & in F.T.A. 7: 505 (1898); Polhill in Journ. E. Afr. Nat. Hist. Soc. 24: 14, fig. 13 (1962); E.P.A. 1536 (1971); Hanid in U.K.W.F.: 684 (1974); Blundell, Wild Fl. E. Afr., reprint: 422, fig. 147 (1992). Lectotype, chosen by Cufod. (1971): Somalia, Wadaba, *Lort-Phillips & Cole* (K, lecto.!)

C. baudi-candeanum Chiov. in Ann. Bot. Roma 9: 146 (1911); Cufod., Miss. Biol. Borana, Racc. Bot.: 310 (1939) & E.P.A.: 1533 (1971). Type: Ethiopa, Ogadene, Gherar to Amaden, *Baudi & Candeo* (FT, holo.!)

Anthericum undulatifolium Poelln. in F.R. 51: 69 (1942). Type: Kenya, Teita District, Voi, towards Bura Mt., *Engler* 1957 (B, holo.!)

Chlorophytum pauciflorum Poelln. in Portug. Acta Biol., sér. B, 1: 322 (1946); E.P.A.: 1535 (1971). Type: Somalia, Arbarone, *Ellenbeck* 2214 (B, holo.!)

C. boranense Chiov. in Webbia 8: 15, fig. 4 (1951); E.P.A.: 1533 (1971). Type: Ethiopia, Sidamo, *Corradi* 4629 (FT, holo.!)

C. tertalense Chiov. in Webbia 8: 18, fig. 6 (1951); E.P.A.: 1536 (1971). Type: Ethiopia, Sidamo, El Banno, Tertale, *Corradi* 4662 (FT, holo.!)

20. **C. tuberosum** (*Roxb.*) *Baker* in J.L.S. 15: 332 (1876); P.O.A. C: 140 (1895); Baker in F.T.A. 7: 508 (1898); F.P.S. 3: 268 (1956); Polhill in Journ. E. Afr. Nat. Hist. Soc. 24: 15, fig. 15 (1962); F.W.T.A., ed. 2, 3: 102 (1968); E.P.A.: 1537 (1971); Nordal & Thulin in Nordic Journ. Bot. 13: 277 (1993); U.K.W.F., ed. 2: 318 (1994); Thulin in Fl. Somalia 4: 47 (1995); Nordal in Fl. Ethiopia 6: 97 (1997). Type: India; t. 138 in Roxb., Pl. Coromandel 2 (1800)

Plants 20–50(–100) cm. high. Rhizome short, irregular, often bearing fibrous remnants from previous years' leaves; roots swollen, with robust dark tubers up to 7 cm. long distally. Leaves rosulate, glabrous, lanceolate, 10–50 cm. long, 1–3(–6) cm. wide, sheathing below, apex acute to obtuse, sometimes cataphylls with ciliate margin. Peduncle stout, 10–35 cm. long, leafless, glabrous, terete. Inflorescence unbranched or rarely with a few basal branches, racemose, dense, up to 17 cm. long; floral bracts hyaline, lanceolate up to 15 mm. Pedicels articulated in lower half to near the middle, up to 10 mm. long, 2–3 at the lower node. Flowers large and showy, sweetly scented, shallowly bowl-shaped; tepals ± 20 mm. long and 6–9 mm. wide, 10–14-veined (the only species in the genus with more than 7-veined tepals). Stamens shorter than the tepals; filaments filiform, 5–6 mm. long; anthers versatile, 3–6 mm. long. Style straight, ± as long as the stamens. Capsule deeply triquetrous, 15 mm. long, 8–12 mm. wide, slightly emarginate. Seeds irregularly folded, 2 mm. in diameter.

UGANDA. Karamoja District: Lolachat, Apr. 1956, *J. Wilson* 229! & Matheniko, Nadunget R., Apr. 1948, *Wabin* 57!; Acholi District: Paloga, Apr, 1943, *Purseglove* 1356!

KENYA. Northern Frontier Province: Dandu, 10 Apr. 1952, *Gillett* 12751!; Meru District: Meru National Park, Kiulo R., May 1972, *Ament & Magogo* 127!; Kitui District: 5 km. NW. of Mutomo, 11 Nov. 1965, *Gillett* 16957!

FIG. 6. *CHLOROPHYTUM SOMALIENSE* — **1**, habit, × $^1/_2$; **2**, flower, x $1^1/_2$; **3**, young fruits, × $1^1/_2$. All from cultivated specimen of *Nordal* 2286. Drawn by Annegi Eide.

TANZANIA. Mbulu District: Tarangire National Park, 23 Nov. 1969, *Richards* 24746!; Pare District: Same, 7 km. SSW. of railway station, 4 Apr. 1971, *Wingfield* 1337!; Dodoma District: Kazikazi, 4 Dec. 1931, *B.D. Burtt* 3524!; Iringa District: Msembe–Mbagi, 17 Dec. 1970, *Greenway & Kanuri* 14962!

DISTR. **U** 1; **K** 1, 2, 4, 6, 7; **T** 1–3, 5, 7; Nigeria, Cameroon, Sudan, Eritrea, Ethiopia, Somalia and India.

HAB. In woodland or bushland, often in degraded vegetation, usually on badly drained, heavy black cotton soils, sometimes on lighter, sandy and lateritic soils, often in seasonally flooded areas; 30–1700 m.

SYN. *Anthericum tuberosum* Roxb., Pl. Coromandel 2: 20, t. 138 (1800)
A. ornithogaloides A. Rich., Tent. Fl. Abyss 2: 332 (1850). Type: Ethiopia, near Tchelatchekanne, Tacazze R. valley, *Quartin-Dillon* (P, lecto.!)
Chlorophytum russii Chiov., Result. Sci. Miss. Stef.-Paoli, Coll. Bot.: 172 (1916); E.P.A.: 1536 (1971). Lectotype, chosen by Nordal & Thulin (1993): Somalia, Giumbo, Juba [Giuba] plains, *Paoli* 275 (FT, lecto.!)
Anthericum kilimandscharicum Poelln. in F.R. 51: 72 (1942). Type: Tanzania, Kilimanjaro, near Moshi, *Peter* 52273 (B, holo.!)
Chlorophytum kulsii Cufod. in Senck. Biol. 50: 243, fig. 3 (1969) & E.P.A.: 1534 (1971). Type: Ethiopa, Gamo-Gofa, Sagan plains E. of Konso, *Kuls* 401 (FR, holo.!)

21. **C. zavattarii** (*Cufod.*) *Nordal* in Nordic Journ. Bot. 13: 65 (1993); Nordal & Thulin in Nordic Journ. Bot. 13: 278 (1993); U.K.W.F., ed. 2: 318 (1994); Thulin, Fl. Somalia 4: 48 (1995); Nordal in Fl. Ethiopia 6: 97 (1997). Type: Ethiopia, Borana, Moyale, *Cufodontis* 689 (FT, holo.!)

Plants 30–60 cm. Rhizome short, bearing fibrous remnants from previous years' leaves; roots swollen and thick throughout their length. Leaves rosulate, broadly lanceolate, ± rounded at apex, 10–25 cm. long, 2.5–5 cm. wide, sheathing below, often with a hyaline margin; sheathing cataphylls sometimes present. Peduncle terete, 10–25 cm., leafless, but sometimes with one sterile bract below the inflorescence. Inflorescence a much-branched panicle; floral bracts short, lanceolate and acute, supporting stiffly patent lateral branches up to 30 mm. long. Pedicels articulated near the apex, stiffly patent to semipatent, up to 25 mm. long, 2–3-nate at the lower nodes. Tepals patent, 5–6 mm. long, 2–2.5 mm. wide, white with a greenish dorsal stripe, 3-veined. Stamens shorter than the tepals; filaments filform, scabrid, sometimes pinkish in upper parts, ± 3 mm. long; anthers 1.5–2 mm. long. Style straight. Capsule deltate in cross-section, emarginate, ± 4 mm. long, Seeds saucer-shaped, 2 mm. in diameter.

KENYA. Northern Frontier Province: Moyale, 17 Oct. 1952, *Gillett* 14060! & Mt. Kulal, *Herlocker* 194!; Meru District: Isiolo–Wajir, 6 km. E. of junction with Marsabit road, 7 Dec. 1977, *Stannard & M.G. Gilbert* 804!

DISTR. **K** 1, 4; Ethiopia and Somalia

HAB. In open woodland or bushland, often on rocky outcrops, dominated by *Acacia, Commiphora* and/or *Combretum*, often in degraded and grazed vegetation, on red lateritic soils or on white gritty soils, over granite or limestone; 1050–1600 m.

SYN. *Anthericum zavattarii* Cufod, Miss. Biol. Borana, Racc. Bot.: 308, fig. 100 (1939); Polhill in Journ. E. Afr. Nat. Hist. Soc. 24: 9 (1962); E.P.A.: 1532 (1971)

22. **C. cameronii** (*Baker*) *Kativu* in Nordic Journ. Bot. 13: 62 (1993); Nordal & Thulin in Nordic Journ. Bot. 13: 262 (1993); U.K.W.F., ed. 2: 316, t. 145 (1994), as '*cameroonii*'; Nordal in Fl. Ethiopia 6: 102 (1997). Type: Tanzania, Kigoma District, S. of Kawele, *Cameron* (K, holo.!)

Plants tufted, 10–100 cm. high. Rhizome short, moniliform, horizontal, covered with fibrous remains of old leaf-bases; roots thin, wiry, bearing distinct distant tubers. Leaves subdistichous to distichous, the innermost sheathing below, folded or flat, broadly linear to lanceolate, glabrous (sometimes papillate, rarely with ciliate margins), sometimes petiolate, glabrous, 10–80 cm. long, 0.5–4 cm. wide, midribs (and sometimes veins) prominent, sometimes papillate, cataphylls and outer leaf-bases sometimes with reddish-brown spots or stripes, sometimes cataphylls gradually turning to falcate leaves. Peduncle flat, winged, glabrous to slightly papillate, 10–100 cm. long. Inflorescence usually unbranched (sometimes a few basal branches), open with elongated internodes, up to 25 cm. long; rhachis sinuate, winged in lower part, terete above, glabrous; floral bracts small, lanceolate, acute, sometimes purplish, up to 12 mm. long. Pedicels articulated below or close to the middle (sometimes close to the base), 4–12 mm. long in fruit, 2–4(–7)-nate at the lower nodes. Tepals patent, 10–15(–20) mm. long, 4–7 mm. wide, (3–)5–7-veined, whitish, outer ones usually pinkish or brownish-pinkish (sometimes greenish) striped on the outside, inner ones white, all turning reddish when withering. Stamens declinate, shorter than the perianth, arranged in two groups, usually 4 in the upper and 2 in the lower; filaments glabrous, 2–6 mm., shorter than the 4–7 mm. long anthers. Style declinate, slightly exserted. Capsule obovoid, rounded-deltate in cross-section, transversely ridged, emarginate, 6–8.5 mm. long, covered by perianth-remnants. Seeds irregularly folded, 1–2 mm. in diameter.

Cataphylls and outer leaf-bases with reddish brown spots or stripes . . . . a. var. *cameronii*
Cataphylls and outer leaf-bases green, without reddish brown spots or stripes:
  Leaves glabrous on the lamina (rarely ciliate on margins); peduncle glabrous . . . . b. var. *pterocaulon*
  Leaves with ciliate margins and ± pubescent on lamina; peduncle ciliate . . . . c. var. *grantii*

a. var. **cameronii**

Cataphylls and outer leaf-bases with reddish-brown spots or stripes.

UGANDA. Karamoja District: Lodoketemit [Lodoketeminit], 20 May 1963, *Kerfoot* 4948!; Acholi District: Kitgum Matidi, Apr. 1943, *Purseglove* 1508!; Masaka District: near Kirumba, 26 Oct. 1969, *Lye & Rwaburindore* 4588!

KENYA. Trans-Nzoia District: SW. Elgon, Saboti [Seboti] Hill, 11 May 1958, *Symes* 373!; Masai District: Masai Mara Reserve, Keekorok [Egerok], 19 Sept. 1947, *Bally* 5415! & Keekorok Lodge, Mara R., 22 Dec. 1969, *Greenway* 13890!

TANZANIA. Musoma District: near Nyaroboro [Nyabora] range, 30 Jan. 1963, *Greenway & Harvey* 10924!; Dodoma District: Rungwa R. Game Reserve, Musa, 28 Feb. 1963, *Mdelwa* 22!; Kondoa District: Great North Road, 24 km. S. of Kondoa, 19 Jan. 1962, *Polhill & Paulo* 1217!

DISTR. **U** 1–4; **K** 2–6; **T** 1, 2, 4–8; Congo, Zaire, Rwanda, Sudan, Ethiopia, Angola, Zambia and Malawi

HAB. In woodland with tall grasses, dry mixed bushland/thornscrub, in dry tall grassland, mountain grassland or on rocky outcrops, often on well-drained sandy or gravelly shallow red soils, sometimes on swampy seasonally wet (mbuga) black clayish soils; 650–2650 m.

SYN. *Anthericum cameronii* Baker in J.L.S. 15: 314 (1876); P.O.A. C: 139 (1895); Hanid in U.K.W.F.: 676 (1974); Blundell, Wild Fl. E. Afr., reprint: 418, fig. 140 (1992)

A. uyuiense Rendle in J.L.S. 30: 415 (1895); P.O.A. C: 139 (1895); Baker in F.T.A. 7: 485 (1898); F.P.S. 3: 263 (1956); Polhill in Journ. E. Afr. Nat. Hist. Soc. 24: 8, fig. 7 (1962); F.W.T.A., ed. 2, 3: 97 (1968); F.P.U., ed. 2: 206 (1971). Type: Tanzania, between Uyui and the coast, 1886, *W.E. Taylor* (BM, holo.!)

A. congolense De Wild & T. Durand, Ann. Mus. Congo, Bot., sér. 2, 1: 60 (1899). Type: Zaire, around Lubunda, *Dewèvre* 10029 (BR, holo.!)

A. brunneum Poelln. in F.R. 51: 113 (1942). Type: Tanzania, Ufipa District, Kasanga [Bismarckburg], Msagena R., *van Wagenheim* 21 (B, holo.!)

A. brunneum Poelln. var. *angustatum* Poelln. in F.R. 51: 113 (1942). Type: Tanzania, Tabora District, Ugunda, Kakoma, *Boehm* 28 (B, holo.!)

A. nguluense Poelln. in F.R. 51: 117 (1942). Types: Tanzania, Tabora District, Ngulu, E. of Goweko, *Peter* 34846 & Dodoma District, Uyansi, from Chaya [Tschaya] E. to Kazikazi, *Peter* 34312 (B, syn.!)

b. var. **pterocaulon** (*Baker*) *Nordal*, stat. & comb. nov. Type: Angola, Pungo Andongo, *Welwitsch* 3795 (BM, holo.!)

Leaves glabrous on the lamina (rarely with ciliate margins); cataphylls and outer leaf-bases green. Peduncle glabrous. Fig. 1/5–8.

UGANDA. Karamoja District: 80 km. W. of Iriri, 16 May 1971, *Brown* 2225!; Bunyoro District: Bujenje, Biso [Biiso], 15 Apr. 1972, *Synnott* 888!; Mengo District: 16 km. NW. of Nakasongola, 8 Sept. 1955, *Langdale-Brown* 1501!

KENYA. Northern Frontier Province: 15 km. N. of Maralal on road to Baragoi, 26 Oct. 1978, *M.G. Gilbert et al.* 5147!; Nairobi District: Thika Road House, 5 Apr. 1951, *Verdcourt* 461!; Kwale District: between Samburu and Mackinnon Road, near Taru, 3 Sept. 1953, *Drummond & Hemsley* 4156!; Kilifi District: Marafa, *Polhill & Paulo* 834!

TANZANIA. Ufipa District: Mbizi Forest, 26 Nov. 1958, *Napper* 987!; Dodoma District: 20 km. from Itigi on Manyoni road, 25 Apr. 1962, *Polhill & Paulo* 2171!; Iringa District: Sao Hill, Ipogoro–Mkawa track, 12 Dec. 1961, *Richards* 15570!; Lindi District: Lake Lutamba, 14 Jan. 1935, *Schlieben* 5874!

DISTR. **U** 1–4; **K** 1–7; **T** 1–8; Cameroon, Congo, Central African Republic, Zaire, Rwanda, Burundi, Sudan, Ethiopia, Somalia, Angola, Zambia, Malawi, Mozambique and Zimbabwe

HAB. In open woodland, thicket, grassland or rocky outcrops, in seasonally wet areas and in swampy places, on shallow sandy soils or on black clay soil; 0–3300 m.

SYN. *Anthericum zanguebaricum* Baker in J.L.S. 15: 302 (1876) & in F.T.A. 7: 483 (1898); Polhill in Journ. E. Afr. Nat. Hist. Soc. 24: 9 (1962). Type: Kenya, Mombasa, Sept. 1873, *Kirk* (K, holo.!)

A. pterocaulon Baker in Trans. Linn. Soc., Bot., ser. 2, 1: 258 (1878) & in F.T.A. 7: 481 (1898); F.W.T.A., ed. 2, 3: 96 (1968)

A. rubellum Baker in Trans. Linn. Soc., Bot., ser. 2, 2: 352 (1887) & in F.T.A. 7: 482 (1898). Type: Tanzania, Kilimanjaro, Oct. 1884, *Johnston* (K, holo.!)

A. speciosum Rendle in J.L.S. 30: 413, t. 33 (1895); P.O.A. C: 139 (1895); Baker in F.T.A. 7: 486 (1898); W.F.K.: 117 (1948). Type: Tanzania, between Uyui and the coast, Feb. 1887, *W.E. Taylor* (BM, holo.!)

A. giryamae Rendle in J.L.S. 30: 412 (1895); P.O.A. C: 139 (1895); Baker in F.T.A. 7: 487 (1898). Type: Kenya, Kwale District, between Giriama [Giryama] and Shimba Hills, *W.E. Taylor* (BM, holo.!)

A. purpuratum Rendle in J.L.S. 30: 413 (1895); P.O.A. C: 139 (1895); Baker in F.T.A. 7: 485 (1898). Type: Kenya, Teita Hills, Ngurunga, Kifamiko, 3 Apr. 1893, *Gregory* (BM, holo.!)

A. buchananii Baker in F.T.A. 7: 486 (1898). Type: Malawi, Shire Highlands, *Buchanan* 201 (K, holo.!)

A. calateifolium Chiov., Fl. Somala 2: 425 (1932); E.P.A.: 1533 (1971). Type: Somalia, Dalleri, *Senni* 218 (FT, holo.!)

A. purpuratum Rendle var. *longipedicellatum* Poelln. in F.R. 51: 73 (1942). Type: Tanzania, Tabora District, Ngulu, Malongwe, *Peter* 34577 (B, holo.!)

A. kyimbilense Poelln. in F.R. 51: 80 (1942). Type: Tanzania, Rungwe District, Kyimbila, *Stolz* 2358 (B, holo.!, BR!, K, iso.)

A. subpapillosum Poelln. in F.R. 53: 133 (1944); Polhill in Journ. E. Afr. Nat. Hist. Soc. 24: 8, fig. 6 (1962). Type: Kenya, Nairobi District, *Scott Elliot* 162 (B, holo.!, K, iso.!)

A. sp. near *pterocaulon* 1 sensu Polhill in Journ. E. Afr. Nat. Hist. Soc. 24: 7 (1962)

A. sp. near *pterocaulon* 2 sensu Polhill in Journ. E. Afr. Nat. Hist. Soc. 24: 7 (1962)

[*A. cooperi* sensu Hanid in U.K.W.F.: 676 (1974); Cribb & Leedal, Mount. Fl. S. Tanz.: 184 (1992), *non* Baker]

Chlorophytum pterocaulon (Baker) Kativu in Nordic Journ. Bot. 13: 63 (1993)

C. zanguebaricum (Baker) Nordal in Nordic Journ. Bot. 13: 65 (1993); Nordal & Thulin in Nordic Journ. Bot. 13: 278 (1993); U.K.W.F., ed. 2: 317, t. 144 (1994); Thulin, Fl. Somalia 4: 49 (1995).

c. var. **grantii** (*Baker*) *Nordal*, comb. & stat. nov. Type: Tanzania, Tabora District, Tura, *Speke & Grant* (K, holo.!)

Leaves with ciliate margins and ± pubescent on lamina; cataphylls and outer leaf-bases green. Peduncle ciliate.

UGANDA. Ankole District: Mpororo, Kirere, *Stuhlmann*
KENYA. Machakos, 1952, *Kirrika* 141! & 20 Dec. 1931, *van Someren* 1588!; Teita District: Taveta, *Johnston*!
TANZANIA. Mwanza District: Kalemera, 20 Apr. 1953, *Tanner* 1344! (small form); Dodoma District: Mgunda Mkali, 28 Dec. 1860, *Grant*!; Iringa District: 56 km. N. of Iringa, 4 Feb. 1962, *Polhill & Paulo* 1342!
DISTR. **U** ?2; **K** 4, 7; **T** 1, 4, 5, 7; not known elsewhere
HAB. Woodland, in grass on red sandy soil, among rocks; 650–1500 m.

SYN. *Anthericum grantii* Baker in Trans. Linn. Soc. 29: 160, t. 103B (1875) & in F.T.A. 7: 488 (1898)
A. venulosum Baker in Trans. Linn. Soc., Bot., ser. 2, 2: 352 (1887) & in F.T.A. 7: 488 (1898); Polhill in Journ. E. Afr. Nat. Hist. Soc. 24: 8 (1962); Hanid in U.K.W.F.: 678 (1974). Type: Kenya, Teita District, Taveta, *Johnston* (K, holo.!)
?*A. stuhlmannii* Engl., P.O.A. C: 138 (1895); Baker in F.T.A. 7: 488 (1898). Type: Uganda, Ankole District, Mpororo, Kirere, *Stuhlmann* (B, holo.)
A. grantii Baker var. *muenzneri* Engl. & K. Krause in E.J. 45: 128 (1910). Type: Tanzania, Ufipa District, Mtembwa–Ebene, *Münzner* 127 (B, holo.!)
A. muenzneri (Engl. & K. Krause) Poelln. in F.R. 51: 74 (1942)
Chlorophytum grantii (Baker) Nordal in Nordic Journ. Bot. 13: 63 (1993)

NOTE. *C. cameronii* s.l. is a variable species complex as to size, general robustness, leaf width, coloration and indumentum. Several attempts have been made to define more homogenous species within the complex. The variation pattern is, however, reticulate and difficult to accommodate taxonomically. We have chosen to keep three taxa at varietal level based on leaf coloration and indumentum, as they are usually easily distinguished (although transitional forms are sometimes met). More local studies on variation patterns are needed.

Two names have often been used wrongly for representatives of the complex in East Africa, viz. *Anthericum cooperi* Baker (e.g. Hanid in U.K.W.F) and *Anthericum anceps* Baker. Both are here regarded as southern taxa, not reaching East Africa. The former, now transferred to *Chlorophytum cooperi* (Baker) Nordal, is a taxon with 3-veined tepals and scabrid filaments, much longer than the anthers, with a northern limit of distribution in South Africa. The latter, now transferred to *Chlorophytum anceps* (Baker) Kativu, is a small-flowered taxon (tepals up to 8 mm.) with a branched inflorescence and with a northern limit of distribution in Zimbabwe.

23. **C. goetzei** *Engl.* in E.J. 28: 361 (1900). Type: Tanzania, Iringa District, Uhehe, *Goetze* 526 (B, holo.!)

Plants tufted, glabrous or hairy, 20–40 cm. high. Rhizome small, moniliform, bearing fibrous remains of old leaf-bases; roots fibrous, many, with tubers near the tips. Leaves distichous or nearly so, firm, filiform, clasping below, ciliate, 10–35 cm. long, 1–2 mm. wide. Peduncle terete, ribbed. Inflorescence unbranched; floral bracts small, dark brown, apiculate. Pedicels articulated below the middle, 2–6 mm. long, 2–3-nate at the lower nodes. Perianth white; tepals 7–10 mm. long, 5-veined, green, brown or rarely pinkish on the keel. Stamens declinate, probably in 2+4 arrangement; filaments shorter than the anthers. Style declinate, as long as the stamens. Capsule shallowly lobed, 2–5 mm. long, transversely ridged. Seeds irregularly folded, 1–1.5 mm. in diameter.

TANZANIA. Kigoma District: Uvinza, W. of Malagarasi, *Peter* 36031! & N. of Lugufu, Kigamba Mwagao, *Peter* 36684!; Iringa District: Uhehe, *Goetze* 526!
DISTR. **T** 4, 7; not known elsewhere
HAB. In open woodland and grassland; ± 1000 m.

SYN. *Anthericum gramineum* Poelln. in F.R. 51: 25 (1942). Type: Tanzania, Kigoma District, N. of Lugufu, Kigamba Mwagao, *Peter* 36684 (B, holo.!)
A. remotiflorum Poelln. F.R. 51: 114 (1942). Type: Tanzania, Kigoma District, Uvinza, ± 1084 km. on Dar es Salaam–Tabora railway, W. of Malagarasi, *Peter* 36031 (B, holo.!)
A. goetzei (Engl.) Poelln. in F.R. 53: 136 (1944)

NOTE. Only known from the three types cited above. The specimens are depauperate, and may belong to the mainly southern tropical *C. galpinii* (Baker) Kativu complex from which they differ by the filiform leaves and the 5 tepal-veins. More material is needed.

24. **C. paucinervatum** (*Poelln.*) *Nordal* in Nordic Journ. Bot. 13: 63 (1993). Type: Tanzania, Lindi District, Tendaguru, *Schlieben* 6011 (B, holo.!)

Plants up to 60 cm. high. Rhizome small, moniliform, with fibrous remains of old leaf-bases; roots fibrous, probably with tubers (not seen). Leaves grass-like, narrowly linear, rosulate, with prominent red-brownish bands at the base especially on outer leaves, narrowly linear, ciliate on the margins, sometimes also with scattered hairs on the lamina, up to 55 cm. long, 4 mm. wide, often narrower. Peduncle compressed, narrowly winged, glabrous, up to 40 cm. long. Inflorescence simple, sometimes with basal branches, open with elongated internodes; rhachis terete, glabrous; floral bracts small, ovate, ciliate, outer ones shortly awned. Pedicels articulated near or below the middle, up to 8 mm. long in fruit, 2–3-nate at the lower nodes. Perianth open, star-shaped; tepals white to pinkish, ± 12 mm. long, ± 3 mm. wide, 5-veined. Stamens shorter than tepals; anthers 6 mm. long, slightly longer than the filaments. Capsule subglobose, shallowly 3-lobed in cross-section, transversely ridged, ± 5 mm. long, 5 mm. wide. Seeds small, irregularly folded, 1–2 mm. in diameter.

TANZANIA. Lindi District: Tendaguru, 29 Apr. 1930, *Migeod* 685!; Tunduru District: 96 km. from Masasi, 18 Mar. 1963, *Richards* 17910!; Masasi District: 8 km. NE. of Masasi, Masasi Hills, 15 Mar. 1991, *Bidgood, Abdallah & Vollesen* 2009!

DISTR. **T** 8; Malawi and Mozambique

HAB. Shallow peaty soils and wet depressions on granite outcrops, sometimes in woodland and wooded grassland; 100–900 m.

SYN. *Anthericum paucinervatum* Poelln. in F.R. 51: 78 (1942)

NOTE. The species is related to the *C. cameronii* complex, but seems distinct enough to deserve specific rank.

25. **C. sphacelatum** (*Baker*) *Kativu* in Nordic Journ. Bot. 13: 64 (1993). Type: Angola, 112 km. from Ambriz, *Monteiro* (K, holo.!)

Plants tufted or solitary, 35–150 cm. high. Rhizome horizontal, moniliform, densely covered with fibrous remains of old leaf-bases; roots fibrous, bearing tubers distally. Leaves distichous, bases clasping the peduncle, outer ones short and falcate, inner ones gradually longer and more erect, firm, narrowly lanceolate, acute, closely ribbed, densely pubescent or glabrous, scabrid or ciliate on the margins, 18–100 cm. long, 0.5–4 cm. wide. Peduncle mainly leafless, but with a few bract-like leaves just below the inflorescence, sharply angled or compressed and winged, glabrous, ciliate or pubescent. Inflorescence unbranched or with 1–2 condensed branches at the base, ± compact, elongated to subglobose, up to 12 cm. long; floral bracts dark brownish, ovate to broadly lanceolate, aristate, ciliate or glabrous on the margins. Pedicels articulated below the middle, 8–15 mm. long and arcuate in fruit, 2–3-nate at the lower nodes. Perianth white; tepals 9–11 mm. long, 5–9-veined, outer ones with dark blackish-red tips, scabrid on the margins. Stamens shorter than the perianth; filaments dilated at the middle, shorter to as long as the anthers, glabrous. Style declinate, exserted. Capsule obovoid, shallowly 3-lobed to rounded-deltate in cross-section, transversely ridged, 6–8 mm. long, 5–6 mm. wide. Seeds irregularly folded, 1–2 mm. in diameter.

Plants robust, 80–165 cm. high; inflorescence subglobose, dense; peduncle flat and winged, completely glabrous ..... c. var. *hockii*

Plants more slender, up to 90 cm. high; inflorescence moderately dense; peduncle subterete and densely pubescent or flat with ciliate wings, rarely glabrous:

Peduncles subterete, sharply angled above, densely pubescent from the base . . . . a. var. *sphacelatum*
Peduncle flat and winged, glabrous, scabrid or shortly ciliate on the wings . . . . b. var. *milanjianum*

a. var. **sphacelatum**

Plants up to 90 cm. high. Peduncles subterete, sharply angled above, densely pubescent from the base. Inflorescence moderately dense.

TANZANIA. Kilimanjaro, near Moshi, *Braun* 3986!; Kigoma District: Uvinsa, *Peter* 36387!
DISTR. **T** 2, 4; Zaire, Angola, Zambia and Malawi
HAB. In open woodland and wooded grassland, often near seasonal pans; ± 900 m.

SYN. *Anthericum sphacelatum* Baker in J.L.S. 15: 303 (1876) & in F.T.A. 7: 489 (1898)
A. lukiense De Wild. in B.J.B.B. 3: 273 (1911). Type: Zaire, Luki, *Brixhe* 6 (BR, holo.!)
A. lukiense De Wild. var. *intermedium* De Wild. in B.J.B.B. 3: 273 (1911). Types: Zaire, Mayombe, 1909, *Deleval* (BR, syn.!) & Kisantu, 1909, *Gillet* (BR, syn.!)
A. lukiense De Wild. var. *kionzoense* De Wild. in B.J.B.B. 3: 273 (1911). Type: Zaire, Kionzo region, *Gillet* 4001, (BR, holo.!)
A. velutinum De Wild. in F.R. 11: 508 (1913). Type: Zaire, Lubumbashi [Elisabethville], Dec. 1911, *Hock* (BR, holo.!)
A. homblei De Wild. in F.R. 13: 108 (1914). Type: Zaire, Shaba, Kapanda valley, *Homblé* 967 (BR, holo.!)
A. robustulum Poelln. in F.R. 51: 124 (1942). Type: Tanzania, Kigoma District, Uvinsa, *Peter* 36387 (B, holo.!)
A. robustulum Poelln. var. *angustum* Poelln. in F.R. 51: 125 (1942). Type: Tanzania, Kilimanjaro, near Moshi, *Braun* 3986 (B, holo.!)

b. var. **milanjianum** (*Rendle*), *Nordal*, stat. nov. Type: Malawi, Mt. Mulanje, *Whyte* (BM, holo.!)

Plants up to 90 cm. high. Peduncle flat and winged, glabrous, scabrid or shortly ciliate on the wings. Inflorescence moderately dense.

TANZANIA. Ufipa District: Kasanga, 31 Mar. 1959, *Richards* 11025!; Iringa District: Msembe–Mbagi track, 20 Mar. 1970, *Greenway & Kanuri* 13996!; Songea District: 1.5 km. S. of Gumbiro, 27 Jan. 1956, *Milne-Redhead & Taylor* 8556!
DISTR. **T** 2, 4–8; Malawi, Zimbabwe, Botswana and Namibia
HAB. In open woodland, bushland or grassland, on grey to brownish, waterlogged clay soils, sometimes also on stony soils and on termite mounds; 750–1600 m.

SYN: *Anthericum milanjianum* Rendle in Trans. Linn. Soc., Bot., ser. 2, 4: 51 (1894); P.O.A. C: 139 (1895); Baker in F.T.A. 7: 486 (1898)
A. whytei Baker in K.B. 1897: 285 (1897) & in F.T.A. 7: 493 (1898). Type: Malawi, Mt. Zomba, *Whyte* (K, holo.!)
?*A. usseramense* Baker in F.T.A. 7: 486 (1898). Type: Tanzania, Uzaramo District, Bunha, *Stuhlmann* 7014 (B, holo., K, iso. (fragment)!)
A. friesii Weim. in Bot. Notis. 1937: 422, fig. 1 (1937). Type: Zimbabwe, Nyanga, near Cheshire, *Norlindh & Weimarck* 4330 (LD, holo.!, BM!, BR!, PRE, SRGH!, iso.)
A. princeae Engl. & K. Krause var. *latifolium* Poelln. in F.R. 51: 134 (1942). Type: Tanzania, Dodoma District, Uyanzi, *Peter* 34277 (B, holo.!)
Chlorophytum sphacelatum (Baker) Kativu subsp. *milanjianum* (Rendle) Kativu in Nordic Journ. Bot. 13: 64 (1993)

c. var. **hockii** (*De Wild.*) *Nordal*, stat. nov. Type: Zaire, Shaba, Luembe valley, 1910, *Hock* (BR, holo.!)

Plants robust, 80–165 cm. high. Peduncle flat and winged, completely glabrous. Inflorescence subglobose, dense.

TANZANIA. Mbeya District: Mbelezi–Mbozi road, 29 Mar. 1932, *St. Clair-Thompson* 1063!; Songea District: ± 2.5 km. SW. of Kitai, by Nakawali R., 7 Mar. 1956, *Milne-Redhead & Taylor* 9046!

DISTR. **T** 7, 8; Zaire, Angola, Zambia and Malawi
HAB. In open *Brachystegia* woodland, in wet areas and near water courses, on sandy soils or red loams; 900–1750 m.

SYN. *Anthericum hockii* De Wild. in B.J.B.B. 3: 265 (1911)
Chlorophytum sphacelatum (Baker) Kativu subsp. *hockii* (De Wild.) Kativu in Nordic Journ. Bot. 13: 64 (1993)

NOTE. *C. sphacelatum* s.l. is a variable species complex as to size, general robustness, leaf-width and indumentum. Several attempts have been made to define more homogenous species within the complex. The variation pattern is, however, reticulate and difficult to accommodate taxonomically. We have chosen to keep three taxa at varietal level based on general robustness and indumentum characters, as they are usually easily distinguished (although transitional forms are sometimes met). Kativu & Nordal in Nordic Journ. Bot. (1993) referred the same taxa to subspecific rank. The morphological and geographical patterns as now known, indicate that the varietal level is more appropriate for the forms. On herbarium labels and sometimes also in the literature plants belonging in the *C. sphacelatum* and in the *C. cameronii* complexes have been mixed. The decisive identification characters (flower colours and stamen symmetry) may be difficult to interpret on dried material. Material from Kenya (**K** 7), e.g. *Irwin* 502, *Polhill & Paulo* 837 & *Battiscombe* 771, possibly belong in the *C. sphacelatum* complex. More local studies on variation patterns is needed.

26. **C. rubribracteatum** (*De Wild.*) *Kativu* in Nordic. Journ. Bot. 13: 64 (1993). Type: Zaire, Lubumbashi [Elisabethville], Nov. 1911, *Hock* (BR, holo.!)

Plants tufted, 10–40 cm. high. Rhizome short, horizontal, moniliform, covered with fibrous remains of old leaf-bases; roots many, fibrous, slender, bearing tubers at the tips. Leaves distichous, grass-like, linear, ciliate, rarely minutely pubescent, often with a purplish-pink tinge, up to 30 cm. long, 2–4 mm. wide; margins minutely scabrid to ciliate near the base; cataphylls with purple coloration, shortly ciliate on margins and veins, occasionally glabrous. Peduncle terete, often purplish, glabrous, seldom minutely pubescent, up to 20 cm. long. Inflorescence unbranched, with elongated internodes; floral bracts lanceolate, red, aristate, glabrous, up to 10 mm. long, ciliate or glabrous on the margins. Pedicels articulated below the middle, often pinkish, up to 9 mm. long in fruit, 2-nate at the lower nodes. Perianth open, star-shaped; tepals 10–13 mm. long, 3–5 mm. wide, (3–)5–7-veined, outer ones pinkish red, inner ones white with the keels pinkish. Stamens with radial symmetry, shorter than the perianth; filaments terete, 2–3 mm. long, equal to or shorter than the anthers. Style declinate, exserted; stigma pinkish. Capsule reddish, shallowly 3-lobed in cross-section, transversely ridged, emarginate, 6–8.5 mm. long. Seeds irregularly folded, ± 1 mm. in diameter.

TANZANIA. Ufipa District: Chapota, 4 Dec. 1949, *Bullock* 2019!; Chunya District: Lupa Forest Reserve, 17 Feb. 1963, *Boaler* 846!; Songea District: by Mkurira R., 26 Dec. 1955, *Milne-Redhead & Taylor* 7895!
DISTR. **T** 4, 7, 8; Zaire, Burundi, Zambia, Malawi and Mozambique
HAB. In *Brachystegia* woodland and grassland margins, on rocky outcrops and in seasonally flooded grassland, often on sandy, sometimes on clayish, soils; 200–1850 m.

SYN. *Anthericum rubribracteatum* De Wild. in F.R. 11: 508 (1913)
A. roseum Poelln. in F.R. 51: 114 (1942). Type: Tanzania, Songea District, E. of Ungoni, Mampyui, *Busse* 727 (B, holo.!)

NOTE. This species belongs close to the *C. sphacelatum* complex, but is easily recognized by the strong coloration of all parts of the plant.

27. **C. subpetiolatum** (*Baker*) *Kativu* in Nordic Journ. Bot. 13: 64 (1993); Nordal & Thulin in Nordic Journ. Bot. 13: 276 (1993); U.K.W.F., ed. 2: 317, t. 145 (1994); Thulin, Fl. Somalia 4: 48 (1995); Nordal in Fl. Ethiopia 6: 100, fig. 190.7 (1997). Type: Mozambique, Lower Zambezi R., Morambala Mts., *Kirk* (K, holo.!)

Plants variable, clumped or single, 10–40 cm. high. Rhizome thick, horizontal to vertical, distinctly moniliform, occasionally tuberous, bearing fibrous remains of old leaf-bases; roots spongy, swollen at the base, tapering towards the tips. Leaves often dimorphic, distichous or nearly so, ± clasping the base of the peduncle, linear-lanceolate to lanceolate, glabrous or pubescent, sometimes ribbed with raised margin and ribs, 5–30 cm. long, 0.5–2(–5) cm. wide, outer ones wider and cataphyll-like, inner ones petiolate; margins scabrid or ciliate. Peduncles papillate to pubescent particularly in the upper half, occasionally pubescent throughout, 5–40 cm. long. Inflorescence a ± lax raceme, 5–15 cm. long; floral bracts oblanceolate, scabrid to ciliate on the margins, lower ones up to 2 cm. long. Pedicels usually articulated below the middle, 3–6 mm. long, 2–3-nate at the lower nodes. Perianth open, star-shaped; tepals 8–15 mm. long, 2.5–4 mm. wide, 3(–5)-veined, white, sometimes tinged greenish or brownish medially towards the tip outside. Stamens shorter than the tepals; filaments filiform, 2.5–5 mm. long, sometimes broadened towards the base, papillate; anthers, 3.5–5 mm., often somewhat longer than the filaments, twisted when dry. Style declinate. Capsule shallowly 3-lobed to rounded-deltate in cross-section, smooth to slightly ridged, 5–9 mm. long and 6–8 mm. wide, slightly emarginate. Seeds irregularly folded, ± 1.5–2 mm. in diameter. Fig. 7.

UGANDA. Acholi District: Imatong Mts., 3–5 km. SE. of Lotuturu, 17 Feb. 1969, *Lye & Lester* 2111!; Ankole District: Mbarara, 29 Sept. 1970, *Katende* 568!; Mubende District: Kakumiro, 17 May 1945, *A.S. Thomas* 4120!

KENYA. Trans-Nzoia District: slopes of Elgon, above Endebess, 30 Apr. 1961, *Polhill* 398! & S. Cherangani, 15 Feb. 1958, *Symes* 284!; Machakos District: 6 km. N. of Machakos, near Mitituni Market, 20 Nov. 1977, *Gillett* 21642!

TANZANIA. Musoma District: SW. of Klein's Camp, 23 Dec. 1969, *Greenway* 13892!; Kondoa District: Great North Road, Kolo, 12 Jan. 1962, *Polhill & Paulo* 1144!; Mbeya District: Magangwe airstrip, 10 Dec. 1970, *Greenway & Kanuri* 14738!; Songea District: ± 19 km. W. of Songea, near Likuyu R., 31 Dec. 1955, *Milne-Redhead & Taylor* 7953!

DISTR. **U** 1–4; **K** 2–7; **T** 1–8; widespread in the Sudano-Zambesian region from Nigeria east to Ethiopia and south to Angola, Mozambique and Zimbabwe

HAB. In all kinds of open woodland, wooded grassland, grassland, even on ± barren soils; tolerates heavy grazing and erosion, often found on very shallow soil overlying rock, different substrates from sandy to heavy seasonally flooded soils, most common on the latter; 50–2450 m.

SYN. *Anthericum subpetiolatum* Baker in J.L.S. 15: 302 (1876) & in F.T.A. 7: 481 (1898); F.P.S. 3: 263 (1956); F.W.T.A., ed. 2, 3: 96 (1968); E.P.A.: 1532 (1971); F.P.U., ed. 2: 206, fig. 135 (1971); Hanid in U.K.W.F.: 678 (1974); Blundell, Wild Fl. E. Afr., reprint: 419, fig. 142 (1992); Cribb & Leedal, Mount. Fl. S. Tanz.: 184 (1992)

A. monophyllum Baker in J.B. 16: 324 (1878) & in F.T.A. 7: 485 (1898); F.P.S. 3: 261 (1956); E.P.A.: 1531 (1971); Hanid in U.K.W.F.: 678 (1974). Type: Sudan, Jur [Djur], Seriba Ghattas, *Schweinfurth* 1793 (K, holo.!)

?*Chlorophytum polyrhizon* Baker in Gard. Chron., ser. 2, 10: 396 (1878). Type: flowered at Kew, July 1878 from Tanzania, coastal [Zanzibar], *Kirk* (type not preserved)

C. moniliforme Rendle in J.L.S. 30: 418 (1895); P.O.A. C: 139 (1895); Baker in F.T.A. 7: 500 (1898). Type: Tanzania, Tabora District, Unyamwezi, Uyui, Feb. 1887, *W.E. Taylor* (BM, holo.!)

C. fusiforme Rendle in J.L.S. 30: 419 (1895); P.O.A. C: 140 (1895); Baker in F.T.A. 7: 495 (1898). Type: Tanzania, Tabora District, Unyamwezi, Uyui, 3 Feb. 1887, *W.E. Taylor* (BM, holo.!)

Anthericum bragae Engl., P.O.A. C: 138 (1895); Baker in F.T.A. 7: 484 (1898). Type: Mozambique, Beira, *Braga* (B, holo.!)

A. triphyllum Baker in F.T.A. 7: 485 (1898). Type: Sudan, Jur, Jur [Djur] R., *Schweinfurth* 1581 (K, holo.!)

Chlorophytum unyikense Engl. in E.J. 30: 272 (1901). Type: Tanzania, Mbeya District, Unyika, Mbosi Mt., *Goetze* 1426 (B, holo.!)

C. ruwenzoriense Rendle in J.L.S. 38: 239 (1908). Type: Uganda, Mt. Ruwenzori, *Wollaston* (BM, holo.!)

Anthericum stolzii Engl. & K. Krause in E.J. 45: 126 (1910). Type: Tanzania, Rungwe District, Kyimbila, Namulapi, *Stolz* 118 (B, holo.!, K, iso.)

FIG. 7. *CHLOROPHYTUM SUBPETIOLATUM* — **1**, habit, × $^2/_5$; **2**, flower, × ± 3; **3**, capsule, × 5; **4**, seed, × 10. 1, from *Milne-Redhead* 3076; 2, from photo. (*Kativu* 261,0); 3, 4, from *Pawek* 3078. Drawn by Eleanor Catherine.

A. claessensii De Wild. in B.J.B.B. 3: 272 (1911). Type: Zaire, Bokala, *Claessens* 119 (BR, holo.!)
A. malchairii De Wild. in B.J.B.B. 3: 274 (1911). Type: Zaire, Likimi, *Malchair* 320 (BR, holo.!)
A. hecqii De Wild. in B.J.B.B. 3: 277 (1911). Type: Zaire, Lake Tanganyika region, 1899, *Hecq* (BR, holo.!)
A. tuberosum De Wild. in F.R. 11: 508 (1913), *nom. illegit.*, *non* Roxb. Type: Zaire, Shaba, Kambove, Oct. 1911, *Hock* (BR, holo.!)
A. laurentii De Wild. var. *minor* De Wild., Pl. Bequaert. 1: 10 (1921). Type: Zaire, Kabare, *Bequaert* 5503 (BR, holo.!)
A. useguahense Poelln. in F.R. 51: 25 (1942). Type: Tanzania, Handeni/Morogoro District, Useguha, between Kanga and Mlembule, *Peter* 7292 (B, holo.!)
A. ugogoense Poelln. in F.R. 51: 30 (1942). Type: Tanzania, Dodoma District, Ugogo, W. of Dodoma, *Peter* 33161 (B, holo.!)
A. unifolium Poelln. in F.R. 51: 78 (1942). Type: Burundi, Usumbura, *Keil* 201 (B, holo.!)
A. porense Poelln. in F.R. 51: 115 (1942). Type: Tanzania, Kilwa District, Matandu [Mandandu], *Busse* 523 (B, holo.!, EA, iso.!)
A. brehmerianum Poelln. in F.R. 51: 116 (1942). Type: Tanzania, Uzaramo District, Magogoni to Ruvu [Ruwu], *Brehmer* 365 (B, holo.!)
?*A. brehmerianum* Poelln. var. *longibracteatum* Poelln. in F.R. 51: 116 (1942). Type: Tanzania, Lushoto District, Umba Steppe, N. of Mashewa, *Peter* 13530 (B, holo.)
A. bifoliatum Poelln. in F.R. 51: 121 (1942). Type: Tanzania, Songea District, Ungoni, *Busse* 869 (B, holo.!)
?*A. diversifolium* Poelln. in F.R. 51: 122 (1942). Type: Tanzania, Singida District, Isansu, *Kohl-Larsen* 2791 (B,holo.)
A. bussei Poelln. in F.R. 51: 127 (1942). Type: Tanzania, Kilwa District, Donde, Kwa Mpanda, *Busse* 1293 (B, holo.!)
A. lacustre Poelln. in F.R. 51: 128 (1942). Type: Tanzania, Mwanza District, Lake Victoria, Ukerewe I., *Conrads* 298 (B, holo.!)
A. urumburense Poelln. in F.R. 51: 130 (1942). Type: Burundi, Usumbura, *Keil* 76 (B, holo.!)
A. setiferum Poelln. in F.R. 51: 131 (1942). Type: Tanzania, Lindi District, Lutamba, *Schlieben* 5692 (B, holo.!, BR, iso.!)
A. dimorphum Poelln. in Bol. Soc. Brot., sér. 2, 16: 43 (1942). Type: Malawi, Chambo R., *McClounie* 101 (B, holo.!)
A. montanum Poelln. in F.R. 53: 135 (1944). Type: Tanzania, Rungwe District, Kyimbila, *Stolz* 2403 (B, holo.!, BR, K, iso.!)
Chlorophytum buarense Poelln. in Ber. Deutsch. Bot. Ges. 61: 205 (1944). Type: Cameroon, Kongola, Buea [Buar], *Mildebraed* 9416 (B, holo.!)
C. hispidiflorum Poelln. in Portug. Acta Biol., sér. B, 1: 294 (1946). Type: Tanzania, Dodoma District, Turu, from Itigi E. to Bangayega, *Peter* 33732 (B, holo.!)
C. hispidoscapum Poelln. in Portug. Acta Biol., sér. B, 1: 295 (1946). Type: Tanzania, Dodoma District, Turu, from Itigi E. to Bangayega, *Peter* 33859 (B, holo.!)
C. longiantheratum Poelln. in Portug. Acta Biol., sér. B, 1: 302 (1946). Type: Tanzania, Mahenge, *Schlieben* 1519 (B, holo.!)
C. subhispidum Poelln. in Portug. Acta Biol., sér. B, 1: 335 (1946). Type: Tanzania, Dodoma District, Turu, from Itigi E. to Bangayega, *Peter* 33716 (B, holo.!)
C. subhispidum Poelln. var. *glabrum* Poelln. in Portug. Acta Biol., sér. B, 1: 336 (1946). Type: Tanzania, Ugogo, S. of Dodoma, *Peter* 33212 (B, holo.!)
C. subulatum Poelln. in Portug. Acta Biol., sér. B, 1: 337 (1946). Type: Tanzania, Kigoma District, Ujiji, *Peter* 52275 (B, holo.!)
Debesia minor (De Wild.) Robyns & Tournay in F.P.N.A. 3: 352 (1955)
Anthericum sp. sensu Polhill in Journ. E. Afr. Nat. Hist. Soc. 24: 7, fig. 4 (1962). Based on *Chlorophytum moniliforme* Rendle

NOTE. *C. subpetiolatum* is a widespread and very variable species having been described under a great number of different names. It is characterized particularly by the thick moniliform rhizome and the fusiform roots, the papillate to pubescent peduncle and the relatively large flowers. Hanid in U.K.W.F. (1974) subdivided the complex into two species: *Anthericum subpetiolatum* and *A. monophyllum*, based on the degree of leaf-clasping and different leaf-texture and whether the moniliform rhizomes are horizontal or vertical. It might be possible to give varietal status to these forms, but transitional forms occur, and we also suspect strongly that the habitat may influence these growth forms. One form (e.g. *Gillett* 13199 from **K** 1, *Greenway* 13894 from **T** 1, and *Milne-Redhead & Taylor* 7316 from **T** 3) with a robust corm has been identified to *Anthericum brehmerianum* Poelln. (reduced to synonymy above). This form may represent a link to the next species where the corm elongates to a distinct stem. The complex displays different chromosome numbers, but these have not been found to correlate with particular morphological forms.

Chromosome number: 2n=16 (*A. Bjørnstad* 1155 from Tanzania, Iringa District) or 48 (*A. Bjørnstad* 551 from Uganda, Ankole District), cf. Nordal et al. in Mitt. Inst. Allg. Bot. Hamburg 23 (1990).

28. **C. suffruticosum** *Baker* in J.B. 1878: 326 (1878) & in F.T.A. 7: 508 (1898); U.K.W.F., ed. 2: 317 (1994). Type: Kenya, near Mombasa, *Wakefield* (K, holo.!)

Plant with erect (rarely prostrate), eventually branched, aerial stem, up to 50 cm. long, usually ± 2.5 cm. thick, with circular leaf-scars and stiff fibres from old leaf-bases. Roots ± spongy without tubers. Leaves distichous or nearly so, grass-like, born at the apices of the branches, 5–50 cm. long, 1–1.5 cm. wide. Peduncles several in the leaf-axils, 5–40 cm. long. Inflorescence a simple raceme (rarely with some basal branches), 6–20 cm. long; rhachis slightly papillose to pubescent; floral bracts short, up to 2–5 mm. long, boat-shaped, with swollen glandular tissue at the base, extruding a sweet, sticky substance. Pedicels articulated below the middle, 3–6 mm., always solitary at the nodes. Flowers open, white, star-shaped; tepals with a greenish, rarely brownish, median stripe, up to 15 mm. long and 5 mm. wide, 3-veined. Stamens declinate, often arranged in a 5 + 1 arrangement, slightly shorter than the tepals; filaments filiform, ± 4 mm. long, half the length of the anthers. Style declinate. Capsule rounded-deltate in cross-section, with slight transverse ridges, ± 10 mm. long and 8 mm. wide. Seeds irregularly folded to saucer-shaped, 1.5–2 mm. in diameter. Fig. 8.

KENYA. Machakos District: 3 km. from Kyulu station, 10 Jan. 1964, *Verdcourt* 3880!; Mombasa District: Mombasa I., between Government House and golf course, cliff top, 28 Jan. 1953, *Drummond & Hemsley* 1058!; Teita District: Tsavo East National Park, Sobo Rocks, 29 Dec. 1969, *Greenway & Kanuri* 12918! & between Mackinnon Road and Kasigau, Kivuko Hill, 26 Apr. 1963, *Bally* 12675!

TANZANIA. Lushoto District: Soni, 11 May 1953, *Drummond & Hemsley* 2526!; Dodoma District: 5 km. from Kilimatinde–Dodoma road, 16 Apr. 1988, *Bidgood, Mwasumbi & Vollesen* 1157!; Bagamoyo District: Chambezi Agricultural Station, 5 Dec. 1967, *S.A. Robertson* 745!

DISTR. **K** 4, 6, 7; **T** 1–3, 5, 6; not known elsewhere

HAB. Pockets of soil in rocks of granitic or coral outcrops, sometimes on termite mounds; 0–1150 m.

SYN. *C. rhizomatosum* Baker in Gard. Chron., ser. 2, 24: 230 (1885) & in F.T.A. 7: 508 (1898). Type: Tanzania, coastal [Zanzibar], 1884, *Kirk*; flowered at Kew, Aug. 1885 (K, holo.!)

Anthericum acuminatum Rendle in J.L.S. 30: 411 (1895); P.O.A. C: 139 (1895); Baker in F.T.A. 7: 484 (1895). Types: Kenya, Teita Hills, Ngurunga, Kifaniko, 3 Apr. 1893, *Gregory* (BM, syn.!) & Tanzania, between Uyui and the coast, 1886, *W.E. Taylor* (BM, syn.!)

A. campestre Engl., P.O.A. C: 138 (1895); Baker in F.T.A. 7: 483 (1898). Type: Tanzania, Tanga, *Holst* 2103 (B, holo., K, iso.!)

Dasystachys polyphylla Baker in F.T.A. 7: 511 (1898). Type: Tanzania, Kilimanjaro, *Johnston* (K, holo.!)

Anthericum suffruticosum (Baker) Milne-Redh. in K.B. 1936: 489 (1936); Polhill in Journ. E. Afr. Nat. Hist. Soc. 24: 8 (1962); Hanid in U.K.W.F.: 678 (1974); Blundell, Wild Fl. E. Afr., reprint: 419, fig. 143 (1992)

A. inexpectatum Poelln. in F.R. 51: 123 (1942). Type: Tanzania, Lushoto District, between Mkumbara [Mkombara] and Shume [Schume], *Peter* 10934 (B, holo.!)

A. longisetosum Poelln. in F.R. 51: 125 (1942). Type: Tanzania, Morogoro District, Useguha, from Wami [Wame] to Kwadihombo [Kwedihombo], *Peter* 7203 (B, holo.!)

Chlorophytum polyphyllum (Baker) Poelln. in Portug. Acta Biol., sér. B, 1: 363 (1946)

NOTE. A very particular and isolated species, the only one within the genus with a woody stem. In growth form and ecology it tends to resemble representatives of Velloziaceae. Its closest relatives are found within the *C. subpetiolatum* complex, and some robust forms of the latter

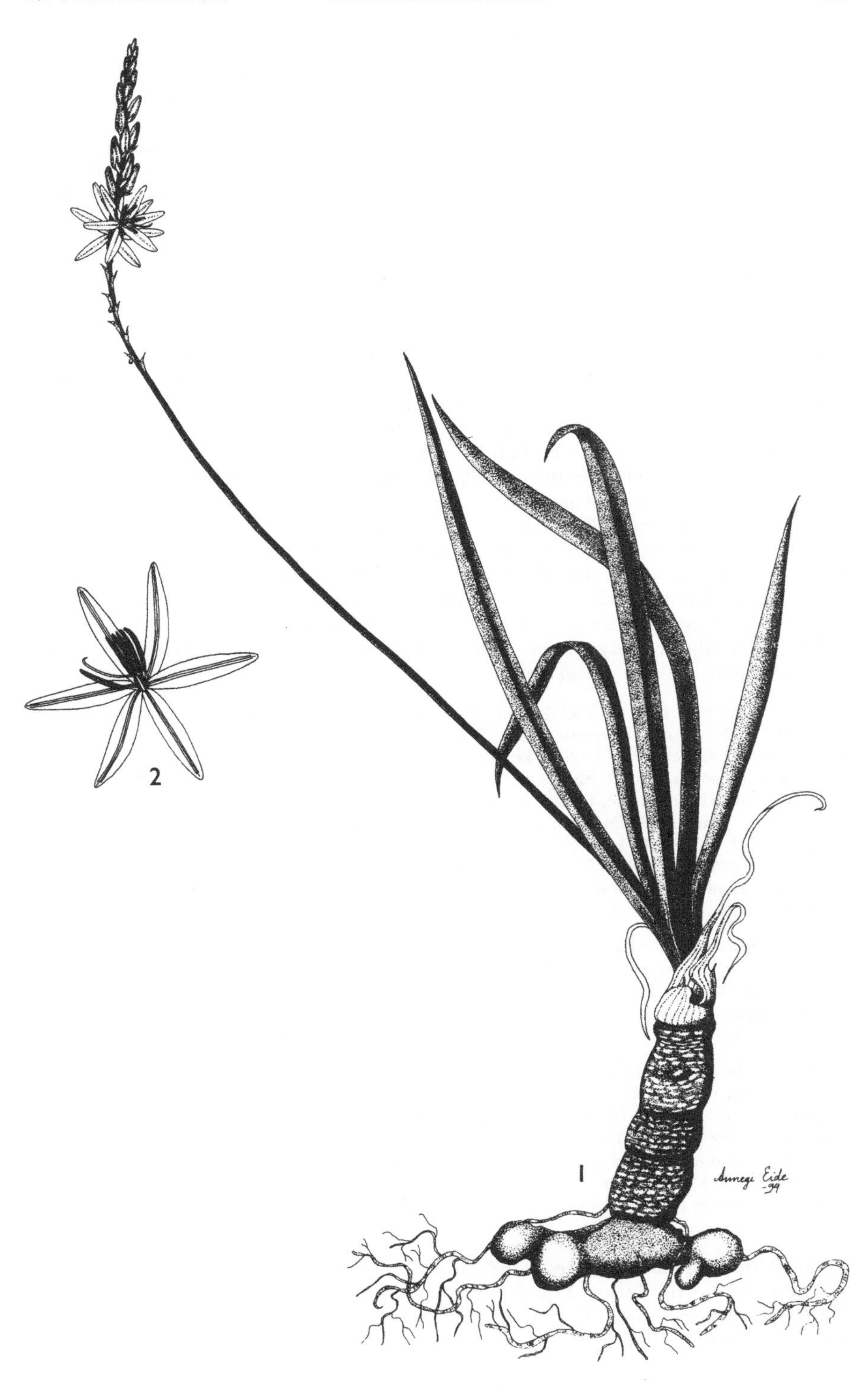

FIG. 8. *CHLOROPHYTUM SUFFRUTICOSUM* — **1**, habit, $^1/_2$; **2**, flower, × $1^1/_2$. All from cultivated specimen of *A. Bjørnstad* 2808. Drawn by Annegi Eide.

(sometimes identified to *Anthericum brehmerianum*) have been mixed with *C. suffruticosum*. Young plants of *C. suffruticosum* lack a well-developed stem and can thus be confused with the *C. subpetiolatum* complex. The former may, however, always be recognized by their flowers solitary at the nodes.

Chromosome number: 2n=32, cf. Nordal et al. in Mitt. Inst. Allg. Bot. Hamburg 23 (1990). The *C. subpetiolatum* complex is shown to include diploid and hexaploid plants. *C. suffruticosum* appears to represent a tetraploid derivative from the complex, derived also in the reduction of the inflorescence to flowers solitary at the nodes.

29. **C. angustissimum** (*Poelln.*) *Nordal*, comb. nov. Type: Tanzania, Tabora District, Ngulu, from Malongwe to Nyahua, *Peter* 34505 (B, holo.)

Plants small, grass-like, clumped, 15–25 cm. high. Rhizome slender, moniliform; roots thin, wiry, with distal tubers 1–2 cm. long. Leaves ± distichous, filiform with expanded bases, often with ciliate margin, up to 35 cm. long, 1–2(–3) mm. wide; cataphylls membranous. Peduncle short, often arcuate below and ± prostrate, without leaves, glabrous, or slightly hairy in lower parts, 1–2(–4) cm. long. Inflorescence a branched panicle, 6–20 cm. long, sometimes prostrate; floral bracts minute, glabrous, up to 3 mm. long. Pedicels articulated at or slightly below the middle, 4–7 mm. long, 2–3-nate at the lower nodes. Perianth open, whitish to yellowish, with green or brownish dorsal stripe or tip; tepals 5–8 mm. long, 2–3 mm. wide, 3-veined. Stamens slightly shorter than the perianth; filaments filiform, ± 4 mm. long and longer than the ± 2.5 mm. long anthers. Style straight. Capsule 2–4 mm. long, somewhat broader than long, covered in remains of the perianth, subglobose, slightly ridged, emarginate, rounded-deltate in cross-section. Seeds saucer-shaped, ± 1 mm. in diameter.

UGANDA. Karamoja District: Lodoketemit [Lodoketeminit], 20 May 1963, *Kerfoot* 4961!; Mbale District: Kaburoron [Kabworon], July 1960, *J. Wilson* 1194!

KENYA. Kilifi District: Arabuko Sokoke, Mida, 1 Jan. 1992, *Luke* 3025!

TANZANIA. Musoma District: Kampi ya Pofu, 28 Feb. 1968, *Greenway & Kanuri* 13348!; Dodoma District: near Kilimatinde, 13 Apr. 1988, *Bidgood, Mwasumbi & Vollesen* 1047!; Njombe District: Ilembula, 9 Jan. 1975, *Brummitt & Polhill* 13659; Songea District: Kwamponjore valley, ± 9.5 km. SW. of Songea, 14 Mar. 1956, *Milne-Redhead & Taylor* 9213!

DISTR. **U** 1, 3; **K** 7; **T** 1–3, 5–8; Burundi, scattered through central Africa to Nigeria

HAB. In shallow pockets of soils on rocky outcrops, sometimes in seasonally wet grassland with shallow soils (dambos); 850–1900 m.

SYN. *Anthericum angustissimum* Poelln. in F.R. 51: 26 (1942)

A. nigericum Hepper in K.B. 22: 459, fig. 5/4–6 (1968) & in F.W.T.A., ed. 2, 3: 97 (1968). Type: Nigeria, Zaria Province, Anara Forest Reserve, *Keay* in *F.H.I.* 37081 (K, holo.!)

Chlorophytum nigericum (Hepper) Nordal in Nordic Journ. Bot. 13: 63 (1993), *non* A. Chev. (1920), *nomen*

NOTE. Specimens belonging to this species have often been confused with the superficially similar, but not closely related *C. calyptrocarpum* (Baker) Kativu. The latter is a southern species, growing from Transvaal and Angola north to central Zambia. It differs particularly by its spongy roots, its bracteate and glandular peduncle and the frequent occurrence of pseudovivipary. One specimen from Tanzania (Kilwa District, ± 6 km. SSE. of Kingupira, 24 Nov. 1975, 125 m., *Vollesen* 3044!) may, however, represent *C. calyptrocarpum*; more material is needed to solve the problem.

30. **C. inconspicuum** (*Baker*) *Nordal* in Nordic Journ. Bot. 13: 63 (1993); Nordal & Thulin in Nordic Journ. Bot. 13: 265 (1993); Thulin, Fl. Somalia 4: 47 (1995); Nordal in Fl. Ethiopia 6: 97 (1997). Type: Somalia, Ahl Mts. above Mait, *Hildebrandt* 1469 (K holo.!)

Slender plants up to 20 cm. Rhizome short; roots, thin in the proximal parts, swelling to elongated tubers distally. Leaves rosulate, filiform or narrowly linear, 10–25 cm. long, 1–4 mm. wide, glabrous except for the sometimes scabrid margins;

basal sheaths whitish to purplish. Peduncle reduced so that the lower flowers/fruits are found within the leaf-rosette. Inflorescence a simple raceme, rarely branched below, 2–5 cm. long, prostrate; rhachis glabrous to slightly scabrid; floral bracts ovate-acuminate, up to 5 mm. long. Pedicels articulated near or below the middle, glabrous, ± curved, 2–5 mm. long at anthesis, up to 1 cm. in fruit, 2-nate at the lower nodes. Flowers small and inconspicuous, whitish to yellowish with green dorsal bands; tepals 3–5 mm. long, 3-veined. Stamens 3–4 mm.; filaments ± 2.5 mm., glabrous; anthers 0.5–1 mm. long. Capsule deeply triquetrous, large compared to the size of the plant, 8–10 mm. long, 5–6 mm. wide when ripe, with withered perianth at the base. Seeds ± saucer-shaped, 2 mm. in diameter.

KENYA. Northern Frontier Province: Garissa–Hagadera road, 42 km. from Garissa, 27 Nov. 1978, *Brenan, Gillett & Kanuri* 14779! & Modo Gash to Garissa, 74 km. S. of Modo Gash, 13 Dec. 1977, *Stannard & M.G. Gilbert* 1030A!

DISTR. **K** 1; Ethiopia and Somalia

HAB. Dry woodland to degraded bushland, on shallow soils overlying volcanic rocks or limestone, seasonally flooded; ± 250 m.

SYN. *Anthericum inconspicuum* Baker in J.B. 15: 71 (1877) & in F.T.A. 7: 480 (1898); E.P.A.: 1531 (1971)

[*Chlorophytum laxum* sensu Baker in F.T.A.: 503 (1898), pro parte, quoad *Schimper* 2231; E.P.A.: 1534 (1971), *non* R. Br.]

NOTE. The species is recorded from West Africa according to F.W.T.A., but the plants from that area are probably not conspecific with *C. inconspicuum*. The older name, *C. laxum* R. Br., has sometimes been used for this taxon in East Africa. The type of *C. laxum* is from Australia. We have decided to avoid this name, but comparative investigations should be undertaken. Nordal & Thulin in Nordic Journ. Bot. 13: 264 (1993) indicated that a species closely related to *C. inconspicuum*, namely the Somalian species *C. filifolium* Nordal & Thulin might be present on Mafia I. (**T** 6, *Wingfield* 4596!). These species are very similar in general habit, the latter differs particularly by capsule size and shape (3–4 mm. long and somewhat wider than long), and by having smaller flowers and roots without tubers. More material is needed to settle the identity of the Mafia material.

31. **C. polystachys** *Baker* in J.B. 16: 326 (1878) & in F.T.A. 7: 509 (1897), as '*polystachyum*'; F.P.S. 3: 268 (1956); F.W.T.A., ed. 2, 3: 99 (1968). Type: Sudan, Jur [Djur], *Schweinfurth* 1838 (BM, holo.!)

Plants variable, often gregarious, 25–60 cm. high. Rhizome horizontal, moniliform, patterned by concentric scars, often bearing fibrous remains of old leaf-bases; roots wiry, bearing tubers. Leaves distichous, firm, linear, the bases often red-brownish to deep green and with a white band at the top especially in young plants, 25–75 cm. long, 0.6–3 cm. wide, glabrous, the margins minutely scabrid to ciliate. Peduncle sometimes arcuate below, terete, glabrous, 20–40 cm. long, often with 1–3 small bracts below the inflorescence, sometimes bronze-coloured in upper part. Inflorescence a lax panicle, up to 25 cm. long; rhachis angled, glabrous, scabrid, or shortly ciliate on the angles; floral bracts glabrous. Pedicels articulated below the middle, pinkish brown, up to 5 mm. long in fruit, 2-nate at lower nodes. Perianth greenish white; tepals 4–5 mm. long, scabrid, 3–5-veined, sometimes with a brownish stripe outside. Stamens as long as the perianth; anthers shorter than the minutely papillate filaments, 1.5 mm. long. Style filiform, exserted. Capsule subglobose, trigonous, smooth, 3–7 mm. long, 5–9 mm. wide. Seeds disc-shaped, up to 3.5 mm. in diameter.

KENYA. Trans-Nzoia District: near Endebess, slopes of Elgon, 1.5 km. NW. of Suam Saw Mill, 7 May 1954, *Rayner* 546! & NE. Elgon, Apr. 1951, *Tweedie* 905!; N. Nyeri District: 53 km. from Nyahururu [Thomson's Falls] on road to Nanyuki, 4 Apr. 1977, *Hooper & Townsend* 1651!

TANZANIA. Masai District: 7 km. from Kibaya to Kondoa, 16 Jan. 1965, *Leippert* 5465!; Kondoa District: 32 km. S. of Kondoa, 19 Jan. 1962, *Polhill & Paulo* 1224; Iringa District: Great North Road, 6.5 km. N. of Iringa, 10 Mar. 1962, *Polhill & Paulo* 1701!; Songea District: 12 km. W. of Songea, by Kimarampaka stream, 20 Jan. 1956, *Milne-Redhead & Taylor* 8382!

DISTR. **K** 3, 4; **T** 1–5, 7, 8; Cameroon, Sudan, Zambia and Zimbabwe
HAB. In open woodland and on rocky outcrops, stony soils or black clay soil; 150–2900 m.

SYN. *C. asparagiflorum* Engl. in E.J. 28: 361 (1900). Type: Tanzania, Iringa District, Uhehe, Muhinde Steppe, *Goetze* 517 (B, holo.!)
C. palustre Engl. & K. Krause in E.J. 45: 138 (1910). Types: N. Cameroon, Buba, *Ledermann* 4075 & 4129 (B, syn.!) & Ngesik, *Ledermann* 4284 (B, syn.) & Benue to Garoua, *Ledermann* 4467 (B, syn.)
Anthericum giganteum Poelln. in F.R. 51: 28 (1942). Type: Tanzania, Lindi District, Lutamba, *Schlieben* 5843 (B, holo.!, B, iso.!)
Chlorophytum acrothyrsum Poelln. in Portug. Acta. Biol., sér. B, 1: 274 (1946). Type: Tanzania, Tabora District, Unyanyembe, E. of Malongwe, *Peter* 34484 (B, holo.!)
C. holotrichum Poelln. in Portug. Acta Biol., sér. B, 1: 296 (1946). Type: Tanzania, Kilosa District, Ussagara, SW. of Kidete, *Peter* 32813a (B, holo.!)
C. subrugosum Poelln. in Portug. Acta Biol., sér. B, 1: 336 (1946). Type: Tanzania, Dodoma District, between Saranda and Makutupora, *Peter* 33703 (B, holo.!)
C. tuberigenum Poelln. in Portug. Acta Biol., sér. B, 1: 342 (1946). Type: Tanzania, Dodoma District, Uyansi, Lake Chaya, *Peter* 34162 (B., holo.!)

NOTE. *C. polystachys* is a variable species. In the Flora Zambesiaca area, two cytotypes (2n=14 and 2n=28) are recorded. The latter is more robust, often ciliate on the leaf-margins, with an appressed shortly branched inflorescence, shortly ciliate rhachis, and larger capsules. However, these characters are not always constant and intermediates are found. The intraspecific variation should be further studied also in East Africa. The species is closely related to *C. pubiflorum*, differing only by displaying different degrees of pubescence on the pedicels and flowers. The type of *C. holotrichum* has leaves with ciliate leaf-margins and veins, pointing towards *C. vestitum*; in other characters, however, it matches *C. polystachys*. The justification for specific rank for the three species (32, 33 and 34) might be questioned, and more studies should be undertaken particularly in SW. Tanzania.

32. **C. pubiflorum** *Baker* in J.L.S. 15: 329 (1876); P.O.A. C: 140 (1895); Baker in F.T.A. 7: 509 (1898). Type: Mozambique, Zambezi delta, *Kirk* (K, holo.!)

Plants 60–80 cm. high. Rhizome horizontal, moniliform; roots thick, tough, bearing tubers at the tips. Leaves distichous, broadly linear to linear-lanceolate, usually glabrous, often with a dark green to brownish-green band at the base, clasping, 20–30 cm. long, up to 12 mm. wide. Peduncle subterete, glabrous, occasionally with 1–2 sterile bracts close to the inflorescence. Inflorescence lax, ascending; racemes 3–10, each subtended by a bract; rhachis angled, glabrous, sometimes sparsely papillate; floral bracts ovate, small, apiculate. Pedicels articulated below the middle, densely papillate-pubescent, up to 10 mm. long in fruit, 2–4-nate at the lower nodes. Perianth greenish white, papillate-hairy, ± 5 mm. long; tepals 3-veined. Stamens as long as the perianth; filaments filiform, glabrous, much longer than the anthers. Style exserted, slightly declinate. Capsule globose, trigonous, ± 5 mm. long, 5 mm. wide. Seeds disc-shaped, ± 2.5 mm. in diameter.

TANZANIA. Tabora District: Unyamwesi, Kombe, towards Usinge, *Peter* 35612!; Ufipa District: Chilo, 2 Jan. 1968, *Richards* 22847!
DISTR. **T** 4; Zambia, Malawi and Mozambique
HAB. In seasonally wet grassland, in sandy clayish soil; ± 1500 m.

SYN. *C. paniculosum* Poelln. in Portug. Acta Biol., sér. B, 1: 321 (1946). Type: Tanzania, Tabora District, Unyamwesi, Kombe, towards Usinge, *Peter* 35612 (B, holo.!)

NOTE. Together with *C. polystachys* and *C. vestitum* this species constitutes a group of closely related taxa.

33. **C. vestitum** *Baker* in J.L.S. 15: 326 (1878); P.O.A. C: 140 (1895); Baker in F.T.A. 7: 506 (1898). Type: Mozambique, Shupanga, *Kirk* (K, holo.!)

Plants 30–70 cm. high. Rhizome horizontal, narrow, moniliform, bearing fibrous remains of old leaf-bases; roots fibrous, with tubers usually on lateral branches. Leaves distichous or nearly so, grass-like, ciliate to pubescent, linear, usually without coloured banding at the base, sheathing the lower peduncle, 20–50 cm. long, 2–6 mm. wide. Peduncle terete, ± pubescent, rarely glabrous, up to 45 cm. long. Inflorescence a lax panicle; rhachis terete, glabrous; floral bracts small, ovate, acuminate. Pedicels often articulated below the middle, sometimes in upper half, up to 10 mm. long in fruit, 2-nate at the lower nodes. Perianth yellowish green; tepals 3–4 mm. long, ± 1 mm. wide, 3–5-veined. Stamens as long as the perianth; filaments glabrous to papillate, ± 3 mm. long; anthers versatile, ± 1 mm. long. Capsule subglobose, triquetrous, 3–4 mm. long, 4–5.5 mm. wide. Seeds disc-shaped, 2 mm. in diameter.

TANZANIA. Musoma District: Kirawira, 20 Feb. 1968, *Greenway, Kanuri & Turner* 13310!; Ufipa District: Sumbawanga, 30 km. from Kasanga on road to Mbala [Abercorn], 24 Nov. 1959, *Richards* 11827!; Songea District: Kimarampaka, 5 Feb. 1956, *Milne-Redhead & Taylor* 8495!

DISTR. **T** 1, 4, 5, 7, 8; Zaire, Zambia and Malawi

HAB. In open *Brachystegia* woodland and wooded grassland, often in shade; 950–1600 m.

SYN. *C. pilosissimum* Engl. & K. Krause in E.J. 45: 139 (1910). Type: Zaire, Lulemba R., *Kassner* 2436 (B, holo.!, BR, iso.!)

NOTE. Together with *C. polystachys* and *C. pubiflorum* this species constitutes a group of closely related species. *C. vestitum* seems distinct enough to justify specific rank. Some intermediates do, however exist, as noted under *C. polystachys.*

34. **C. gallabatense** *Baker* in J.L.S. 15: 325 (1876) & in F.T.A. 7: 504 (1898); Chiov., Fl. Somala 2: 425 (1932); F.P.S. 3: 267 (1956); Polhill in Journ. E. Afr. Nat. Hist. Soc. 24: 12, fig. 10 (1962); F.W.T.A., ed. 2, 3: 99 (1968); Nordal & Thulin in Nordic Journ. Bot. 13: 265 (1993); U.K.W.F., ed. 2: 318, t. 145 (1994); Thulin, Fl. Somalia 4: 48 (1995); Nordal in Fl. Ethiopia 6: 99 (1997). Lectotype, chosen by Marais & Reilly in K.B. 32: 660 (1978): Sudan/Ethiopian border, Gallabat, Matamma, *Schweinfurth* 10 (K, lecto.!)

Plants 15–60(–75) cm. high. Rhizome horizontal, short, moniliform; roots fairly thin, shortly branched, bearing tubers 2–3 cm. long, mainly on the lateral branches. Leaves rosulate, glabrous, lanceolate, broadest at the middle, narrowed and clasping below, occasionally petiolate, 10–75 long, (1–)2–4(–6) cm. wide, with distinct midrib, the margins glabrous or minutely scabrid, often undulate. Peduncle terete, glabrous, leafless, occasionally with a single sterile bract below the inflorescence, 10–50 cm. long. Inflorescence paniculate; branches usually with short internodes (except for the lowermost ones), more elongated in shade-forms; rhachis glabrous to papillate; floral bracts 5–15(–30) mm. long, 1–3 mm. wide, larger and longer in shade forms. Pedicels articulated near or above the middle, 3–10 mm. long, slightly papillate, 2–4-nate at the lower nodes. Perianth with reflexed tepals, greenish, 5–7 mm. long, 1.5–2 mm. wide, 3(–5)-veined. Stamens as long as the perianth; filaments fusiform, scabrid, ± 3 mm. long; anthers 2–3 mm. long. Style straight, as long as the stamens. Capsule deeply triquetrous, emarginate, 3–6 mm. long, 4–7 mm. wide, smooth. Seeds disc- to saucer-shaped, ± 2.5 mm. in diameter.

UGANDA. Karamoja District: Lodoketemit, 12 July 1958, *Kerfoot* 413!; Teso District: Serere, Apr. 1932, *Chandler* 577!; Mengo District: 7 km. from Kiwoko to Tweyanze, 23 Apr. 1956, *Langdale-Brown* 2076!

KENYA. Northern Frontier Province: Dandu, 30 Mar. 1952, *Gillett* 12659!; Meru District: 19 km. ENE. of Isiolo on road to Mado Gashi, 16 May 1972, *Gillett* 19673!; Masai District: SE. of Kajiado, Enkorika, 24 Nov. 1977, *Kuchar & Mwendwa* 7880!

TANZANIA. Ngara District: Nyakiziba, 27 Apr. 1960, *Tanner* 4904!; Mbeya District: Ruaha National Park, near Magangwe, Isiki, *A. Bjørnstad* 2069!; Lindi District: 9.5 km. S. of Mbemkuru R. on Kilwa to Lindi road, 6 Dec.1955, *Milne-Redhead & Taylor* 7565!

DISTR. **U** 1–4; **K** 1–7; **T** 1, 3–8; widespread in the Sudano-Zambesian region from Senegal east to Somalia and south to Mozambique and Zimbabwe

HAB. In ± degraded and heavily grazed deciduous woodland and bushland, or in disturbed grassland, sometimes in shade, more often in open places, sometimes on termite mounds, on lateritic sandy loam to heavy clay; 100–2300 m.

SYN. *C. ukambense* Baker in F.T.A. 7: 504 (1898); Polhill in Journ. E. Afr. Nat. Hist. Soc. 24: 15 (1962); E.P.A.: 1537 (1971). Type: Kenya, Kitui, *Hildebrandt* 2650 (K, holo.!)

C. rivae Engl. in Ann. R. Ist. Bot. Roma 9: 244 (1902); E.P.A.: 1535 (1971). Type: Ethiopia, Savati, near Sagan R., *Ruspoli & Riva* 1522 (FT, holo.!)

C. ginirense Dammer in E.J. 65 (1905); E.P.A.: 1535 (1971). Type: Ethiopia, Bale, Ginir, *Ellenbeck* 1960 (B, holo.!)

C. ramulosum De Wild. in B.J.B.B. 3: 275 (1911). Type: Zaire, between Bambile and Amadi, *Sereti* 244 (BR, holo.!)

C. breviscapum Dammer in E.J. 48: 362 (1912), *nom. illegit.*, *non* Dalzell (1850). Type: Tanzania, Kilwa District, Matandu [Mandandu], *Busse* 547 (B, holo.!, EA, iso.!)

C. breviflorum De Wild. in F.R. 11: 513 (1913). Type: Zaire, Lubumbashi [Elisabethville], *Homblé* 153 (BR, holo.!)

C. hockii De Wild. in F.R. 11: 514 (1913). Type: Zaire, Lubumbashi [Elisabethville], Dec. 1911, *Hock* (BR, holo.!)

C. bequaertii De Wild., Pl. Bequaert 1: 14 (1921); F.W.T.A., ed. 2, 3: 102 (1968). Type: Zaire, Kabare, *Bequaert* 5410 (BR, holo.!)

C. longiramum Poelln. in Portug. Acta Biol., sér B, 1: 304 (1946). Type: Tanzania, Dodoma District, Turu, between Itigi and Bangayega, *Peter* 33961 (B, holo.!)

C. longistylum Poelln. in Portug. Acta Biol., sér. B, 1: 305 (1946). Type: Tanzania, Iringa District, Uhehe, Lula, *Goetze* 504 (B, holo.!, K, iso.!)

C. polyscapum Poelln. in Portug. Acta Biol., sér. B, 1: 326 (1946). Type: Tanzania, Kigoma District, Ujiji, *Peter* 38975 (B, holo.!, B, iso.!)

C. riparium Poelln. in Portug. Acta Biol., sér. B, 1: 332 (1946). Type: Tanzania, Njombe District, Lupembe, upper stream of Ruhudji [Ruhudje] R., *Schlieben* 391 (B, holo.!, BR, iso.!)

?*C. socialis* Poelln. in Portug. Acta Biol., sér. B, 1: 335 (1946). Type: Tanzania, Ulanga District, between Ulanga and Ebene, Mahenge, *Schlieben* 1528 (B, holo.!, K, iso.!) – depauperate specimen that probably belongs here (or in *C. micranthum*)

C. umbraticolum Poelln. in Portug. Acta Biol., sér. B, 1: 345 (1946). Type: Tanzania, Ulanga District, Mahenge, Mbangala, *Schlieben* 1805 (B, holo.!)

C. elachystanthum Cufod. in Senck. Biol. 50: 242, fig. 2 (1969) & in E.P.A.: 1533 (1971). Type: Ethiopia, Gamo-Gofa, *Kuls* 357 (FR, holo.!)

NOTE. Shade forms superficially resemble *C. comosum* by having semiprostrate inflorescences. They can be distinguished by flower colour (greenish in *C. gallabatense*) and root system (small tubers on lateral branches in *C. gallabatense*). Forest forms with prostrate inflorescences (e.g. *Mbago* 1074, Kigoma District, Gombe Stream and *Bidgood, Sitoni, Vollesen & Whitehouse* 4331, Mpanda District, Livendabe Forest) might deserve taxonomic recognition. Flowering material is needed. For further notes see under *C. micranthum* and *C. floribundum.*

Chromosome number: 2n=14 based on countings from Kenya, cf. Nordal et al. in Mitt. Inst. Allg. Bot. Hamburg 23 (1990).

35. **C. micranthum** *Baker* in J.B. 16: 325 (1878) & in F.T.A. 7: 507 (1898); F.P.S. 3: 268 (1956); Polhill in Journ. E. Afr. Nat. Hist. Soc. 24: 13 (1962); F.P.U., ed. 2: 206 (1971); Nordal & Thulin in Nordic Journ. Bot. 13: 269 (1993); U.K.W.F., ed. 2: 318 (1994); Nordal in Fl. Ethiopia 6: 99 (1997). Type: Sudan, Jur [Djur], *Schweinfurth* 1745 (K, holo.!, B, iso.!)

Plants hysteranthous, up to ± 20 cm. high. Rhizome short, moniliform, with fibrous remains from previous years' leaves; roots fairly narrow, with small tubers on lateral root-branches. Leaves rosulate, glabrous, narrowly lanceolate, never collected when fully developed. Peduncle leafless, up to 10 cm. long. Inflorescence simple or slightly paniculate; rhachis glabrous; floral bracts small. Pedicels articulated near the middle, up to 2 cm. long, 2–4-nate at the lower nodes. Tepals patent to slightly reflexed, greenish, 4–5 mm. long, 3-veined. Stamens as long as the perianth; filaments 3–4 mm. long; anthers shorter. Style filiform, exserted. Capsule triquetrous, ± 4 mm. long. Seeds not known.

UGANDA. Acholi District: near Patiko, 3 km. SW. of Lotuturu, 17 Feb. 1969, *Lye & Lester* 2041! & Madi Opei, Apr. 1943, *Purseglove* 1370!; Karamoja District: Mt. Kadam [Debasien], *Eggeling* 2586!

KENYA. Trans-Nzoia District: S. Cherangani, 15 Feb. 1958, *Symes* 295! & Kitale, 5 Mar. 1953, *Bogdan* 3675!; Mar. 1966, *Tweedie* 3254!; Uasin Gishu District: Kipkarren, Mar. 1932, *Brodhurst Hill* 740!

DIST. **U** 1–3; **K** 3; Burundi, Sudan and Ethiopia

HAB. Woodland or wooded grassland on sandy soils, often burnt; 900–2100 m.

NOTE. This species is closely related to *C. gallabatense* and might just be a small and hysteranthous (fire-adapted?) form, not deserving more than subspecific rank. Fully developed material is needed for further study.

36. **C. floribundum** *Baker* in K.B. 1897: 285 (1897) & in F.T.A. 7: 505 (1898). Type: Malawi, Mt. Zomba, *Whyte* (K, holo.!, B, iso.!)

Plants 16–55 cm. high. Rhizome short, vertical, moniliform; roots narrow, shortly branched, bearing small, rounded tubers on the branches. Leaves rosulate, 4–13 to a plant, lanceolate to broadly lanceolate, clasping at the base, rarely with short petioles, broadest at the middle, narrowed above to a fine point, usually glabrous, 10–40(–60) cm. long, 1–5(–8) cm. wide. Peduncle terete, compressed when dry, glabrous below, often papillate to pubescent above, up to 24 cm. long. Inflorescence exserted, with 2–many racemes, very rarely unbranched, bearing flowers from the base; racemes subtended by lanceolate bracts, these up to 5 cm. long; rhachis terete, often papillate; floral bracts lanceolate, softly apiculate, up to 10 mm. long. Pedicels articulated above the middle, 4–5 mm. long in fruit, papillate, 2–4-nate at the lower nodes. Perianth greenish or whitish, keeled green; tepals 5–8 mm. long, 1–2 mm. wide, 3-veined. Stamens as long as the perianth; filaments fusiform, papillose, ± 5 mm. long; anthers versatile, 1 mm. long. Style straight, ± the length of the stamens. Capsule triquetrous, emarginate, 5–7 mm. long, 6–7 mm. wide. Seeds disc- to saucer-shaped, ± 2 mm. diameter.

TANZANIA. Biharamulo District: Lusahungu, 15 Oct. 1960, *Tanner* 5606A!; Ufipa District: Lake Sundu, 8 Nov. 1958, *Richards* 10251!; Rungwe District: Kyimbila, Madehani, 3 Dec. 1913, *Stolz* 2324a!; Songea District: ± 8 km. W. of Songea, 1 Jan. 1956, *Milne-Redhead & Taylor* 7998!

DISTR. **T** 1, 4, 7, 8; Zambia, Malawi and Zimbabwe

HAB. In open *Acacia-Commiphora* or *Brachystegia* woodland; 800–2150 m.

SYN. *C. puberulum* Engl., P.O.A. C: 139 (1895); Baker in F.T.A. 7: 505 (1898). Type: Tanzania, Biharamulo District, Ukome, *Stuhlmann* 877 (B, holo.!)

C. papillososcapum Poelln. in Portug. Acta Biol., sér. B, 1: 322 (1946). Type: Tanzania, Rungwe District, Kyimbila, Madehani, *Stolz* 2324a (B, holo.!, BM, BR, K, iso.!)

NOTE. *C. floribundum* is closely related to *C. gallabatense* with which it can be confused. The former is characterized by pubescence on upper part of the peduncle, the rhachis and the pedicels, further by somewhat longer capsules. The taxonomic delimitation might be justified only at subspecific rank.

37. **C. humifusum** *Cufod.*, Miss. Biol. Borana, Racc. Bot.: 311, fig. 101 (1939); Polhill in Journ. E. Afr. Nat. Hist. Soc. 24: 13 (1962); E.P.A.: 1534 (1971); Nordal & Thulin in Nordic Journ. Bot. 13: 265 (1993); Nordal in Fl. Ethiopia 6: 100 (1997). Type: Ethiopia, Moyale, *Cufodontis* 690 (FT, holo.!)

Small plants, rarely more than 12 cm. Rhizome moniliform, carrying short fibrous remnants from older leaves; roots wiry, shortly branched, bearing small tubers (up to 1 cm. long) mainly on the lateral root-branches. Leaves ± distichous, glabrous, narrowly lanceolate, petiolate, up to 16 cm. long and 1(–2) cm. wide. Peduncle only 1–2 cm. long, prostrate. Inflorescence, up to 15 cm. long, lying flat on the ground, simple or with one basal branch, lax, with elongated internodes; rhachis scabrid to pubescent; floral bracts up to 5 mm., sometimes ciliate. Pedicels articulated in lower half, 6–9 mm. at anthesis, elongating slightly in fruit, 2-nate at the lower nodes. Perianth open, white; tepals 5–7 mm. long, less than 2 mm. wide, 3-veined. Stamens shorter than the tepals; filaments fusiform, scabrid, 4 mm. long, anthers 1–1.5 mm. long. Style straight, slightly shorter than the stamens. Capsule ± 4 mm. long. Seeds not known.

KENYA. Northern Frontier Province: Mandera, 5 km. E. of junction of Banissa and Derkali roads, 5 May 1978, *M.G. Gilbert & Thulin* 1464! & Dandu, 1 May 1952, *Gillett* 12661! & Moyale, 4 Nov, 1952, *Gillett* 14139!

DISTR. **K** 1; Ethiopia

HAB. In shallow stony soils in bushland or woodland often dominated by *Acacia* and *Commiphora*; 450–850 m.

NOTE. This species has sometimes been confused with slender shade forms of *C. gallabatense*. It is, however, distinct by its distichous and narrower leaves and pure white flowers.

38. **C. zingiberastrum** *Nordal & A.D. Poulsen*, sp. nov. Species habitu cum *C. orchidastrum* Lindley, sed differt radicibus angustioribus non tomentosis; inflorescentia breviore et ramiferiore; bracteis longioribus et latioribus, sed non vaginantibus; foliis et bracteis in sicco flavo-bruneolis, non nigrescentis; pedicellis angulatis vel anguste alatis supra articulum; antheris et capsulis brevioribus. Typus: Malawi, Northern Province, Mzimba District, Mzuzu, Marymount, *Pawek* 7987 (WAG, holo!)

Plants (15–)25–70 cm. high. Rhizome short, vertical, moniliform; roots thin with elongated tubers on lateral rootlets. Cataphylls and petioles sheathing, that of the inner leaf covering large parts of the peduncle, thus forming a pseudostem; lamina lanceolate to oblanceolate, 10–40 cm. long, 3–8 cm. wide, widest just below or at the middle, attenuate, acute to acuminate, glabrous. Peduncle slender, terete, up to 50 cm., glabrous. Inflorescence a lax panicle, 12–18 cm. long, first internode of the branches elongated; branch and flower bracts large and leafy, lower ones up to 7 cm. long, 0.8–1.8 cm. wide, narrowing towards the base and not sheathing the pedicels; both leaves and bracts drying yellowish brown, never blackish. Pedicels 9–10(–14) mm. long, articulated above the middle and angulate, sometimes developing 3 narrow wings between the joint and the flower, 2–4-nate at the nodes. Perianth greenish; tepals 4–6 mm. long, 1–1.5 mm. wide, 3-veined. Filaments fusiform and papillose, 3–4 mm. long; anthers 1.3–2 mm. Style slightly declinate, longer than the tepals at anthesis. Capsule triquetrous, 4–5 mm. long, 5–7 mm. wide, with truncate base and emarginate apex. Seeds disc-shaped, black, papillate, up to 3 mm. in diameter

TANZANIA. Mbeya District: Ruaha National Park, 5 km. W. of Magangwe Ranger Post, 10 May 1972, *A. Bjørnstad* 1723!; Mikindani District: Mtwara–Mikindani road, 13 Mar. 1963, *Richards* 17863!; Masasi District: Ndanda, 10 Mar. 1991, *Bidgood, Abdallah & Vollesen* 1885!

DISTR. **T** 7, 8; Zambia, Malawi and Mozambique

HAB. In *Brachystegia* woodland, on sandy to gritty-clayey soils overlying rocks and on termite mounds; 550–1950 m.

NOTE. This species has up to now been confused with a West to Central African rain-forest taxon, *C. orchidastrum* Lindl. to which it bears a superficial similarity. The latter differs, however, in several traits: it has thicker, more tomentose roots with tubers along the main axes; the leaves dry blackish and the lamina is widest in the lower half with a truncate base; the inflorescence is longer and less branched; the bracts are shorter and embracing the pedicels and buds; the inflorescence displays more flowers per node (up to 7); the pedicels are shorter, rarely more than 8 mm., and terete without angles; the capsules are longer, and the seeds saucer- rather than disc-shaped. More information on this species, including illustration, distribution map and electron microscope scan of the seeds, will be published in K.B. (1998).

39. C. **macrophyllum** (*A. Rich.*) *Aschers.* in Schweinf., Beitr. Fl. Aethiop.: 294 (1867); Baker in J.L.S. 15: 323 (1875) & in F.T.A. 7: 498 (1898); F.P.S. 3: 266 (1956); Polhill in Journ. E. Afr. Nat. Hist. Soc. 24: 13 (1962); F.W.T.A., ed. 2, 3: 99 (1968); E.P.A.: 1535 (1971); Blundell, Wild Fl. E. Afr., reprint: 422, fig. 146 (1992); Nordal & Thulin in Nordic Journ. Bot. 13. 269 (1993); U.K.W.F., ed. 2: 318, t. 145 (1994); Nordal in Fl. Ethiopia 6: 102, fig. 190.8 (1997). Lectotype, chosen by Nordal & Thulin (1993): Ethiopia, Djeladjekanne, *Quartin-Dillon* (P, lecto.!)

Plants 20–80 cm. high, drying yellowish brown to olive-green, very rarely blackish. Rhizome short, compact; roots medium thick, expanding to spindle-shaped tubers up to 3 cm. long. Leaves rosulate, not petiolate nor forming a stem, broadly lanceolate, 10–70 cm. long, 2–10 cm. wide, with undulate or crisped margins, glabrous. Peduncle terete, leafless (or with a few sterile bracts connected to the inflorescence), stout, erect, up to 50 cm. long, glabrous. Inflorescence up to 30 cm., dense, usually unbranched; rhachis glabrous; lower bracts 2.5–6–(13) cm. long, ordinary floral bracts, 2.5–3 cm. long. Pedicels articulated in the upper half, whitish, particularly above the articulation, (8–)10–14 mm. long, 3–5-nate at the lower nodes. Perianth whitish (translucent), turning brownish after anthesis; tepals semi-patent, 8–15 mm. long, 2–4 mm. wide, usually 5-veined. Stamens equal to to shorter than the tepals; filaments fusiform, widest in upper half, scabrid, 3–5 mm. long, somewhat shorter than the 4–6 mm. long anthers. Style declinate. Capsule deltate in cross-section, emarginate, 6–11 mm. long, usually somewhat longer than wide, blackish when dehiscing. Seeds saucer-shaped, ± 2–2.5 mm. in diameter. Fig. 9.

KENYA. Nairobi, corner of Uhuru Highway/Langata Road, 8 May 1975, *Kabuye & Ng'weno* 502!; Masai District: Ngong Hills, near Kiserian, 12 Apr. 1960, *Verdcourt, Hemming & Polhill* 2652! & 34 km. from Narok–Ngorengore, 11 Dec. 1963, *Verdcourt* 3826!

TANZANIA. Masai District: near Lake Lagarja, 1 Jan. 1963, *Greenway & Turner* 10920!; Kondoa District: 24 km. N. of Kolo, 12 Jan. 1962, *Polhill & Paulo* 1143!; Mbeya District: Poroto Mts., 20 Jan. 1961, *Richards* 13990!

DISTR. **K** 4–6; **T** 1–3, 5–7; widespread in tropical Africa from Senegal east to Ethiopia and south to Mozambique and Zimbabwe

HAB. In riverine or dry evergreen forest, usually in shade, but also in more open woodland, bushland and swampy grassland on black soils, rarely on coral rock in coastal areas; 100–2800 m.

SYN. *Anthericum macrophyllum* A. Rich., Tent. Fl. Abyss. 2: 334 (1850)

Chlorophytum fuchsianum De Wild., Ann. Mus. Congo, Bot., sér. 5, 1: 102 (1904). Type: Zaire, Kisantu, *Gillet* 902 (BR, lecto.!, K, isolecto.!)

C. massaicum K. Krause in E.J. 57: 236 (1921). Type: Tanzania, Arusha, *Holtz* 3347 (B, holo.!)

C. kyimbilense Poelln. in F.R. 51: 80 (1942). Type: Tanzania, Rungwe District, Kyimbila, *Stolz* 2358 (B, holo.!, B, iso.!)

C. hyacinthinum Poelln. in Portug. Acta Biol., sér. B, 1: 298 (1946). Type: Tanzania, Arusha District, Meru, Ngare Nanyuki, *Peter* 2215 (B, holo.!)

C. kilimandscharicum Poelln. in Portug. Acta Biol., sér. B, 1: 300 (1946). Type: Tanzania, Kilimanjaro, Moshi, *Merker* 417 (B, holo.!)

C. longipaniculatum Poelln. in Portug. Acta Biol., sér. B, 1: 303 (1946). Type: Tanzania, Kilimanjaro, Kibo, *Endlich* 318 (B, holo.!)

C. macrophyllum (A. Rich.) Aschers. var. *kyimbilense* Poelln. in Portug. Acta Biol., sér. B, 1: 310 (1946). Type: Tanzania, Rungwe District, Kyimbila, *Stolz* 2335 (B, holo.!, K, iso.)

Dasystachys melanocarpa Chiov. in Webbia 8: 20, fig. 7A (1951); E.P.A.: 1538 (1971). Type: Ethiopia, Sidamo, Mega, *Corradi* 4681 (FT holo.!)

Chlorophytum sp. near *macrophyllum* sensu Polhill in Journ. E. Afr. Nat. Hist. Soc. 24: 14, fig. 12 (1962)

C. sp. A sensu Hanid in U.K.W.F.: 685 (1974)

FIG. 9. *CHLOROPHYTUM MACROPHYLLUM* — **1**, habit, × $^1/_2$; **2**, node of inflorescence with flowers in various stages of development, × $^2/_3$. All from cultivated specimen of *Nordal* 1033. Drawn by Annegi Eide.

NOTE. This species has been confused with another broad-leaved species, i.e. the true rain-forest taxon, *C. filipendulum* Baker, from which it differs by having non-petiolate leaves, generally not drying black, longer pedicels, larger flowers, and particularly the relatively long anthers compared to the length of the filament. *C. filipendulum* appears to be more closely linked to the *C. comosum* complex than to *C. macrophyllum*, although intermediate forms exist also between the latter and *C. filipendulum*.

Chromosome number: 2n=28 (tetraploid), based on countings from Ethiopia, cf. Nordal et al. in Mitt. Inst. Allg. Bot. Hamburg 23 (1990).

40. **C. blepharophyllum** *Baker* in J.L.S. 15: 327 (1876); P.O.A. C: 140 (1895); Baker in F.T.A. 7: 501 (1898); Rendle in J.L.S 40: 217 (1911); F.P.S. 3: 267 (1956); Oberm. in Bothalia 7: 711 (1962); Polhill in Journ. E. Afr. Nat. Hist. Soc. 24: 12 (1962); F.W.T.A., ed. 2, 3: 100 (1968); E.P.A.: 1533 (1971); F.P.U., ed. 2: 206 (1971); Nordal & Thulin in Nordic Journ. Bot 13: 262 (1993); U.K.W.F., ed. 2: 317, t. 144 (1994); Nordal in Fl. Ethiopia 6: 105 (1997). Types: Sudan/Ethiopian border, Gallabat, around Matamma, *Schweinfurth* 9 (K, syn.!, P, isosyn.!) & S. Zimbabwe, *Baines* (K, syn.)

Plants growing single or clumped, variable, 10–40 cm. high, drying blackish. Rhizome small, with fibrous remains of old leaf-bases; roots ± spongy, with elongate tubers near the tips. Leaves sometimes slightly hysteranthous, rosulate, olive-green above, paler beneath, lanceolate, moderately firm, clasping the peduncle; lamina glabrous, 10–30 cm. long, ± canaliculate, 1.5–4(–7) cm. wide, margins hyaline and ciliate; cataphylls orange to purplish or with coloured veins, with ciliate and often crisped margins. Peduncle leafless, terete, glabrous, 5–40 cm. long. Inflorescence usually unbranched, occasionally with a few short branches at the base, 8–15 cm. long; rhachis papillate to pubescent; floral bracts linear to lanceolate, lower ones up to 25 mm. long, often shortly ciliate and hairy. Pedicels articulated near to distinctly above the middle, 3–10 mm. long, 2–4-nate at the lower nodes. Perianth whitish tinged greenish cream to brownish, slightly urceolate near the base; tepals ± reflexed, 6–8 mm. long, 1.5–2.5 mm. wide, 3–5-veined, scabrid on margins and veins. Stamens shorter than the perianth; filaments fusiform, scabrid to papillose, 3–5 mm. long, longer than the 1.5–2 mm. long, orange-yellow anthers. Style straight, as long as the stamens. Capsule obovoid, emarginate, triquetrous, 6–10 mm. long, usually longer than broad, with persistent perianth-remnants at the base. Seeds disc-shaped, 2.5–4 mm. in diameter.

UGANDA. Karamoja District: Lonyili Mts., Apr. 1960, *J. Wilson* 930!; Ankole District: 5 km. S. of Mbarara, Ruizi R. valley, 28 Sept. 1969, *Lock* 69/310!; Mengo District: Singo, 31 km. Kiboga Gombolola to Butemba, 25 Mar. 1956, *Langdale-Brown* 2019!

KENYA. Trans-Nzoia District: slopes of Elgon, above Endebess, 30 Apr. 1961, *Polhill* 413!; N. Kavirondo District: N. Nyanza, first bridge on Kakamega road from Uganda–Eldoret road, 5 May 1971, *Mabberley & Tweedie* 1087!

TANZANIA. Ngara District: Bugufi, Mu Rgwanza [Murgwanza], Dec. 1960, *Tanner* 5490!; Morogoro District: 35 km. along Morogoro to Dodoma road, 29 Dec. 1970, *Harris & Kasembe* in *D.S.M.* 2129!; Njombe District: 2 km. SE. Lukumburu, 16 Nov. 1966, *Gillett* 17879!

DISTR. **U** 1–4; **K** 3, 5, 7; **T** 1–8; widespread in the Sudano-Zambesian region from Senegal east to W. Ethiopia and south to Angola, Mozambique and Zimbabwe

HAB. In open deciduous woodland and grassland, on rocky outcrops, often light sandy and stony soils, sometimes grazed and eroded; 50–2450 m.

SYN. *C. ciliatum* Baker in J.B. 16: 325 (1878) & in F.T.A. 7: 505 (1898). Type: Sudan, Jur, Kutchuk Ali, *Schweinfurth* 1521 (K, holo.!)

C. brunneum Baker in F.T.A. 7: 507 (1898). Type: Tanzania, Uzaramo District, near Dar es Salaam, Mboamaji, *Stuhlmann* 6031 (B, holo.!, fragment at K)

C. fibrosum Engl. & K. Krause in E.J. 45: 132 (1910). Types: Cameroon, Lagosche, *Ledermann* 3890 (B, syn.!) & between Duka and Dangadji, *Ledermann* 3647 (B, syn.) & Balda, *Ledermann* 4031 (B, syn.)

C. longebracteatum De Wild. in B.J.B.B. 3: 275 (1911). Type: Zaire, Vankerkhovenville, *Seret* 539bis (BR, holo.!)

C. nigrescens De Wild. in B.J.B.B. 3: 275 (1911). Type: Zaire, Gumbari, Feb. 1906, *Seret* (BR, holo.!)

C. homblei De Wild. in F.R. 11: 514 (1913). Type: Zaire, Lubumbashi [Elisabethville], *Homblé* 153bis (BR, holo.!)
Anthericum viridulobrunneum Poelln. in F.R. 51: 29 (1942). Type: Tanzania, Rungwe District, Kyimbila, Kiwira [Kibila], *Stolz* 1789 (B, holo.!, BR!, K, iso.)
Chlorophytum distachyum Poelln. in Portug. Acta Biol., sér. B, 1: 287 (1946). Type: Tanzania, Tanga District, between Amboni and Gombero, *Peter* 22952 (B, holo.!, B, iso.!)
C. kombense Poelln. in Portug. Acta Biol., sér. B, 1: 302 (1946). Type: Tanzania, Tabora District, Unyamwezi, Kombe, W. towards Usinge, *Peter* 35548 (B, holo.!)
C. saxicolum Poelln. in Portug. Acta Biol., sér. B, 1: 334 (1946). Type: Tanzania, Kigoma District, W. of Uvinza, *Peter* 36371 (B, holo.!)
C. trichocraspedum Poelln. in Portug. Acta. Biol., sér. B, 1: 342 (1946). Type: Tanzania, Kigoma District, Uvinza, near Malagarassi, *Peter* 35913 (B, holo.!)
?*C. viriduliflorum* Poelln. in Portug. Acta. Biol., sér. B, 1: 350 (1946). Type: Tanzania, Dodoma District, Kilimatinde, *Prittwitz* 40 (B, holo.!) – a depauperate specimen which may belong here

NOTE. The species is widely distributed and exhibits several forms, thus the excessive description of species here reduced to synonymy. Chromosome number: 2n=28 (tetraploid), based on countings from Zimbabwe, cf. Nordal et al. in Mitt. Inst. Allg. Bot. Hamburg 23 (1990).

41. **C. amplexicaule** *Baker* in J.L.S. 15: 325 (1876); P.O.A. C: 139 (1895); Baker in F.T.A. 7: 501 (1898). Type: Tanzania, near Lake Tanganyika, *Cameron* (K, holo.!)

Plants 15–35 cm. high, drying blackish. Rhizome short; roots ± spongy, apparently without tubers. Leaves rosulate, broadly lanceolate, strongly clasping the peduncle; lamina glabrous, 6–15 cm. long, 4–7 cm. wide, margins brownish hyaline and ciliate; cataphylls with ciliate and often crisped margins. Peduncle leafless, terete, glabrous, 10–20 cm. long. Inflorescence unbranched or with a few short branches at the base, 8–15 cm. long; floral bracts linear to lanceolate, lower ones up to 25 mm. long, often shortly ciliate and hairy. Pedicels articulated near the middle, up to 4 mm. long, 2–5-nate at the lower nodes. Perianth greenish cream to brownish, slightly urceolate near the base; tepals ± reflexed, 6–8 mm. long, 1.5–2.5 mm. wide, 3–5-veined, patent except for the base. Stamens shorter than the perianth; filaments fusiform, scabrid to papillose, 3–5 mm. long, longer than the 1.5–2 mm. long, orange-yellow anthers. Style straight, as long as the stamens. Capsule not seen.

TANZANIA. Mbeya District: Magangwe airstrip, 10 Dec. 1970, *Greenway & Kanuri* 14763!
DISTR. **T** 4, 7; Zambia
HAB. In open woodland on pale brown sandy loam; ± 1300 m.

NOTE. This taxon may only represent an extremely broad-leaved form of *C. blepharophyllum.*

42. **C. brachystachyum** *Baker* in Gard. Chron., ser. 3, 13: 710 (1893) & in F.T.A. 7: 502 (1898). Type: Malawi, Shire Highlands, *Buchanan*; flowered at Kew, 1893 (K, holo.!)

Plants often clumped, 25–45 cm. high. Rhizome short, vertical, moniliform; roots narrowly spongy to softly fibrous, bearing rounded tubers towards the tips, rarely on lateral branches. Leaves rosulate, oblong-lanceolate, narrowed at the base, attenuate, 5–40 cm. long, 0.5–4 cm. wide, flat, fimbriate and crisped at the (occasionally reddish) ciliate margins. Peduncle terete, as long as the leaves, shallowly angled, glabrous. Inflorescence usually an unbranched, dense raceme, sometimes forked at the base; floral bracts lanceolate, ciliate, abruptly narrowed to a bristle. Pedicels articulated above the middle, ± 5 mm. long in fruit, 2–5-nate at the lower nodes. Perianth white, slightly urceolate; tepals papillate, 4.5–6 mm. long, 1.5–2 mm. wide, 3–5-veined, somewhat reflexed at anthesis. Stamens as long as the perianth; filaments fusiform, papillose, 3–4 mm. long; anthers 1–1.5 mm. long. Ovary papillate; style straight, as long as the stamens. Capsule deltate to slightly 3-lobed in cross-section, emarginate, papillate, tuberculate or transversely lined when mature, ± 4 mm. long, 4–5 mm. wide. Seeds saucer-shaped, ± 2 mm. in diameter. Fig. 10.

FIG. 10. *CHLOROPHYTUM BRACHYSTACHYUM* — **1**, habit, × $^1/_2$; **2**, detail of part of inflorescence, × $1^1/_2$; **3**, capsule, × 8; **4**, seed, × 16. 1, from *Fanshawe* 6085; 2, from photo. (*Adams & Ganda* 624,0); 3, 4, from *Wild* 4748. Drawn by Eleanor Catherine.

TANZANIA. Masai District: Ololmoti [Olomoti], 5 Mar. 1965, *Kapigi* 16!; Dodoma District: Rungwa Game Reserve, Sulangi, 22 Jan. 1969, *V.C. Gilbert* 3546!; Masasi District: 30 km. NW. of Masasi, Chiwale village, 13 Mar. 1991, *Bidgood, Abdallah & Vollesen* 1949!.

DISTR. **T** 2, 4–6, 8; scattered through the Sudano-Zambesian region, from Ghana to the Central African Republic, Angola, Zambia, Malawi, Zimbabwe, Botswana and Namibia

HAB. In areas with rock-like clayey hardpan soils and termite mounds, within open woodland, often near water-courses; 450–1600 m.

SYN. *C. trachycarpum* Oberm. in Bothalia 7: 701, fig. 3 (1962). Type: Namibia, Okavango, between Sambiu and Masari, *de Winter* 4081 (PRE, holo., SRGH, iso.!)

NOTE. Chromosome number: 2n=28 (tetraploid), based on countings from Zimbabwe, cf. Nordal et al. in Mitt. Inst. Allg. Bot. Hamburg 23 (1990).

43. **C. stenopetalum** *Baker* in J.L.S. 15: 331 (1876) & in F.T.A. 7: 502 (1898); F.W.T.A., ed. 2, 3: 100 (1968). Type: Nigeria, Niger R., Nupe, *Barter* (K, holo.!)

Plants 10–30 cm. high. Rhizome short, vertical, moniliform; roots medium thick with distal tubers. Leaves drying blackish, rosulate, glabrous, ± petiolate, lanceolate, clasping at the base, broadest at the middle, 10–40 cm. long, 2–5 cm. wide. Peduncle terete, often with a single bract below the inflorescence, glabrous, 1–3(–6) cm. long. Inflorescence unbranched, occasionally with a single appressed branch at the base, dense, up to 10 cm. long; rhachis subterete, shallowly angled, glabrous; floral bracts lanceolate, glabrous, up to 4 cm. long. Pedicels articulated at the apex, ± 2 mm. long in flower, up to 6 mm. long in fruit, 2–4-nate at the lower nodes. Perianth whitish to greenish; tepals 5–7 mm. long, 3–5-veined. Stamens as long as the perianth; anthers 2.5–3 mm. long; filaments filiform, ± 7 mm. long. Capsule deeply triquetrous, 6–7 mm. long, ± 5.5 mm. wide. Seeds saucer-shaped, ± 1.5 mm. in diameter.

UGANDA. West Nile District: 1 km. SE. of Metu rest camp, 15 Sept. 1953, *Chancellor* 267!; Teso District: Serere, Apr.–May 1932, *Chandler* 605!

TANZANIA. Ulanga District: Mahenge, Mbangala, *Schlieben* 1795!; Iringa District: Kidatu, 8 Mar. 1971, *Mhoro* 683!; Lindi District: E. side of Lake Lutamba, Litipo Forest Reserve, 27 Feb. 1991, *Bidgood, Abdallah & Vollesen* 1743!

DISTR. **U** 1, 3; **T** 6–8; West Africa to Sudan, Angola, Zambia, Malawi and Mozambique

HAB. In open forest and riverine fringes, often on termite mounds; 100–1800 m.

SYN. *C. bracteosum* Baker in Trans. Linn. Soc., Bot., ser. 2, 1: 260 (1878). Type: Angola, Huila, *Welwitsch* 3768 (K, holo.!)

C. menyharthii Baker in F.T.A. 7: 503 (1898). Type: Mozambique, N. Zambezia, Boruma, *Menyharth* 565 (W, holo.!)

C. schweinfurthii Baker in F.T.A. 7: 503 (1898); F.P.S. 3: 267 (1956); E.P.A.: 1536 (1971). Lectotype, chosen by Nordal & Thulin (1993): Sudan, Jur [Djur], Seriba Ghattas, *Schweinfurth* 1968 (B, lecto.!)

C. inarticulatum Poelln. in Ber. Deutsch. Bot. Ges. 61: 128 (1943), Type: Tanzania, Ulanga District, Mahenge, Mbangala, *Schlieben* 1795 (B, holo.!)

NOTE. This species is sometimes difficult to separate from dwarfed forms of *C. macrophyllum* in the herbarium. The latter differs by having articulated pedicels, larger flowers and anthers, different capsule shape and mainly drying yellowish. *C. stenopetalum* as here circumscribed might represent a polyphyletic assemblage of forms with reduced peduncle, rhachis and pedicel above the joint.

44. **C. geophilum** Poelln. in Ber. Deutsch. Bot. Ges. 61: 127 (1943); F.W.T.A., ed. 2, 3: 100 (1968); Nordal & Thulin in Nordic Journ. Bot. 13: 265 (1993); Nordal in Fl. Ethiopia 6: 93 (1997). Type: Tanzania, Tabora District, Unyamwezi, Kombe, *Peter* 35409 (B, holo.!, B, iso.!)

Plants up to 5 cm. high. Rhizome short, carrying narrow roots with elongated tubers. Leaves in a prostrate rosette, oblanceolate, petiolate, obtuse, glabrous, up to 20 cm. long and 4–8 cm. wide, rather thick with prominent veins, occasionally undulate and ciliate. Peduncle terete, up to 2 cm. long, so that the inflorescence appears at ground-level among the leaves. Inflorescence up to 4 cm. long, dense, often branched, sometimes looking almost capitate; floral bracts large, up to 10 mm. long, often ciliate. Pedicels apparently not articulated, ± 5 mm. long, often reflexed in fruit, several at the nodes. Perianth white; tepals 6–8 mm. long, 5-veined; stamens as long as the perianth; anthers ± 2 mm. long, shorter than the filiform filaments. Style filiform, as long as the stamens. Capsule shallowly trigonous, ± 5 mm. long, 4 mm. wide, smooth. Seeds saucer-shaped, ± 2 mm. in diameter. Fig. 11/1.

UGANDA. Mbale District: Sebei, Buligenyi, 19 May 1955, *Norman* 264!

TANZANIA. Iringa District: 30 km. W. of Iringa along road to Idodi, 13 Jan. 1972, *A. Bjørnstad* 1232!; Songea Distict: 32 km. E. of Songea by Mkurira R., 19 Jan. 1956, *Milne-Redhead & Taylor* 8369!; Masasi District: Ndanda mission, 1 Feb. 1991, *Bidgood, Abdallah & Vollesen* 1332!

DISTR. **U** 3; **T** 4, 6–8; Burkina Faso, Ghana, Nigeria, Cameroon, Ethiopia, Zambia and Malawi

HAB. In open *Brachystegia* woodland and grassland, sometimes on termite mounds; 200–1100 m.

SYN. *C. mahengense* Poelln. in Ber. Deutsch. Bot. Ges. 61: 129 (1943). Type: Tanzania, Ulanga District, Mahenge, Mbangala, *Schlieben* 1796 (B, holo.!)

NOTE. The species is very similar to another widespread, ± prostate species with reduced peduncle, *C. pusillum* Baker (1878), with the type from the Sudan. Petiolate leaves which are rather thick with prominent veins and branched inflorescences characterise *C. geophilum*, whereas *C. pusillum* has broad-based thin leaves and unbranched inflorescence. The relation between the two species should be further investigated.

45. **C. pusillum** *Baker* in J.B. 16: 325 (1878) & in F.T.A. 7: 502 (1898); F.P.S. 3: 267 (1956); F.W.T.A., ed. 2, 3: 100 (1968). Type: Sudan, Jur [Djur], *Schweinfurth* 2043 (K, holo.!, B, P, iso.!)

Plants small, often in patches, up to 3.5 cm. high. Rhizome not distinct; roots short, ± spongy, with elongated tubers, roots sometimes reduced to sessile or subsessile elongated tuberous structures. Leaves rosulate, ± prostrate, 1–4 to a plant, rarely more, oblanceolate, thinly membranous, glabrous, (2.5–)5–15 cm. long, (1.5–)3–10 cm. wide; margins crisped, in SW. Tanzania a form occurs with red ciliate margins. Peduncle reduced, up to 1(–3) cm. long. Inflorescence unbranched, dense, 2–5 cm. long. Pedicels apparently without articulation, 1–5 mm. long, several at each node. Perianth white; tepals 4–5 mm. long, ± 1 mm. wide, 3-veined. Stamens as long as the perianth; filaments fusiform and papillate, 2.5–5 mm. long, longer than the anthers. Ovary sessile, with ± 5 ovules per locule; style straight, as long as the stamens. Capsule shallowly deltoid, smooth, ± 4 mm. long, 3 mm. wide. Seeds slightly folded, saucer-shaped, ± 1.5 mm. in diameter. Fig. 11/2–5.

UGANDA. Karamoja District: Labwor, 6 June 1940, *A.S. Thomas* 3721!

TANZANIA. Dodoma District: Rungwa Game Reserve, Sulangi, 10 km. W. of Bagamoyo, 24 Jan. 1969, *V.C. Gilbert* 3627!; Iringa District: 30 km. W. of Iringa, 13 Jan. 1972, *A. Bjørnstad* 1232! & 24 km. E. of Iringa on Morogoro road, 28 Dec. 1966, *Hanid, Hanid & Musumba* 340!

DISTR. **U** 1; **T** 5–7; Senegal, Burkina Faso, Ghana, Nigeria, Cameroon, Central African Republic, Sudan, Zambia and Zimbabwe

HAB. In woodland in shade, often in rock crevices and on termite mounds; 1000–1400 m.

SYN. *C. micans* Engl. & K. Krause in E.J. 45: 135 (1910). Type: N. Cameroon, *Ledermann* 4403 (B, holo.!)

46. **C. comosum** (*Thunb.*) *Jacq.* in Journ. Soc. Imp. Centr. Hort. 8: 345 (1862); Baker in J.L.S. 15: 329 (1876) & in Fl. Cap. 6: 400 (1897); J.M. Wood, Natal Pl. 3: 279 (1902); J.M. Watt & Breyer-Brandwyk, Medic. & Pois. Pl. S. Afr.: 13 (1932). Oberm.

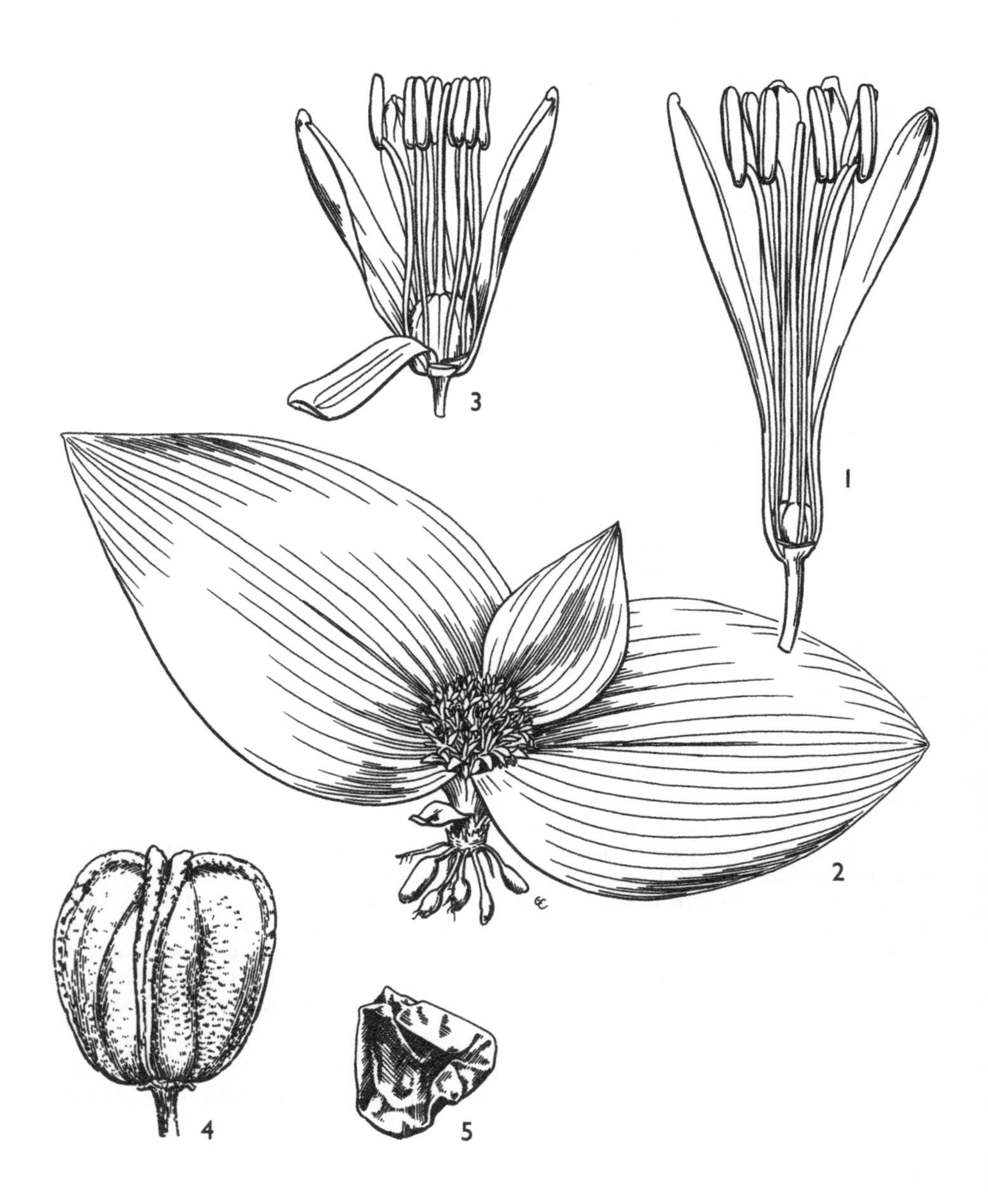

FIG. 11. *CHLOROPHYTUM GEOPHILUM* — **1**, flower with 2 tepals removed, × 6. *CHLOROPHYTUM PUSILLUM* — **2**, habit, × $^{2}/_{3}$; **3**, flower with 2 tepals removed, × 6; **4**, capsule, × 6; **5**, seed, × 12. 1, from *Richards* 512; 2, 3, from *Richards* 13705; 4, 5, from *Fanshawe* 1809. Drawn by Eleanor Catherine.

in Bothalia 7: 698 (1962); Nordal & Thulin in Nordic Journ. Bot. 13: 262 (1993); U.K.W.F., ed. 2: 318 (1994); Nordal in Fl. Ethiopia 6: 102 (1997). Type: South Africa, Cape Province, Uniondale, Langekloof, *Thunberg* (UPS, holo.!)

A variable complex of plants with ± lax leaves and peduncles, sometimes drying blackish. Rhizome short, vertical or sometimes horizontal with elongated internodes; roots ± spongy, often with spindle-shaped tubers. Leaves rosulate, petiolate or not, lanceolate to strap-shaped, usually glabrous, 10–60(–100) cm. long, 1–3(–4) cm. wide, sometimes with scabrid margin. Peduncles often more than one to a plant, 5–20(–50) cm. long, erect to ± drooping, glabrous to scabrid, with a few sterile bracts just below the inflorescence. Inflorescence 10–50(–75 cm.), usually simple, sometimes ± paniculate; rhachis sometimes somewhat scabrid, internode elongation varies considerably giving dense to extremely lax inflorescences; floral bracts 5–20 mm., acute to acuminate. Small plantlets often produced from the bracts of the inflorescence (pseudovivipary). Pedicels articulated in upper half, 4–10 mm. long, slightly reflexed during anthesis, erect in fruit, 2–4-nate at the lower nodes. Perianth slightly urceolate, whitish, often keeled greenish outside and apically; tepals patent to somewhat reflexed at anthesis, 4–7 mm. long, 2–3 mm. wide, 3-veined, with scabrid margins. Stamens exserted; filaments fusiform, dilated above the middle, scabrid to papillose, 3–5 mm. long, longer than the 1–2.5 mm. long, versatile anthers. Style straight, as long as the stamens. Capsule triquetrous to deltate, emarginate, usually somewhat broader than long, 4–9 mm. long, 5–9(–11) mm. wide. Seeds saucer-shaped, slightly folded, 2–3 mm. in diameter.

UGANDA. Karamoja District: Lodoketemit [Lodoketeminit], 20 June 1961, *Kerfoot* 3833!; Toro District: Queen Elizabeth National Park, Mweya peninsula, 7 Oct. 1969, *Lock* 69/327!; Mengo District: Mabira Forest, 1–2 km. E. of Kiwala, 13 Apr. 1969, *Lye, Lester & Morrison* 2488!

KENYA. Northern Frontier Province: Mathews Range, above Mandasion, 7 Dec. 1960, *Kerfoot* 2602!; S. Nyeri District: below Castle Forest Station, 4 Apr. 1970, *Gillett & Mathew* 19097!; Masai District: Masai Mara Game Reserve, Mara R. at the Tanzanian border, 26 Aug. 1976, *Kuchar, Msafiri & Karime* 5064!; Teita District: Ngangao, 8 km. NNE. of Ngerenyi, 15 Sept. 1983, *Drummond & Hemsley* 4335!

TANZANIA. Mwanza District: Saanane I., 17 Jan. 1965, *Carmichael* 1172!; Masai District: Lolkisale, Mar. 1967, *Beesley* 239!; Lushoto District: Mkusu valley, 6.5 km. NE. of Lushoto, 2 Mar. 1963, *Drummond & Hemsley* 1390!; Iringa District: Dabaga Highlands, near Kibengu, Inhangana Forest Reserve, 14 Feb. 1962, *Polhill & Paulo* 1478!

DISTR. **U** 1–4; **K** 1, 3–7; **T** 1–8; a widespread complex species of rain-forest areas of Africa south of the Sahara, from West, Central and East Africa including Ethiopia, southwards to the Cape.

HAB. Undergrowth in rain-forest and riverine forest on brown to black loamy clay, also in crevices in rocks along streams, sometimes epiphytic or penetrating into more open and drier areas (or persists when the forest disappears); sea-level to 2450 m.

SYN. *Anthericum comosum* Thunb., Prodr. Pl. Cap. 1: 63 (1794)

Chlorophytum sparsiflorum Baker in J.L.S. 15: 325 (1876) & in F.T.A. 7: 498 (1898); Polhill in Journ. E. Afr. Nat. Hist. Soc. 24: 13 (1962); F.W.T.A., ed. 2, 3: 100 (1968); E.P.A.: 1536 (1971), as '*sparsifolium*'; Hanid in U.K.W.F.: 684 (1974). Type: Bioko [Fernando Po], *Mann* 388 (K, holo.!)

C. kirkii Baker in Gard. Chron. 17: 108 (1882) & in F.T.A. 7: 506 (1898). Type: Tanzania, coastal [Zanzibar], *Kirk*; flowered at Kew, 1882 (K, holo.!)

C. miserum Rendle in J.L.S. 30: 420 (1895); P.O.A. C: 139 (1895); Baker in F.T.A. 7: 496 (1898). Type: Tanzania, between Uyui and the coast, 1886, *W.E. Taylor* (BM, holo.!)

C. ramiferum Rendle in J.L.S. 30: 421 (1895); P.O.A. C: 139 (1895); Baker in F.T.A. 7: 496 (1898). Type: Tanzania, between Uyui and the coast, 1886, *W.E. Taylor* (BM, holo.!)

C. bukobense Engl., P.O.A. C: 139 (1895); Baker in F.T.A. 7: 504 (1898). Type: Tanzania, Bukoba, *Stuhlmann* 962 (B, holo.!, iso.!)

C. bukobense Engl. var. *kilimandscharicum* Engl., P.O.A. C: 139 (1895); Baker in F.T.A. 7: 504 (1898). Type: Tanzania, Kilimanjaro, Marangu, *Volkens* 1294 (B, holo.!)

C. delagoense Baker in Fl. Cap. 6: 399 (1897). Type: Mozambique, Maputo, Delagoa Bay, *Monteiro* (K, holo.!)

C. gazense Rendle in J.L.S. 40: 216 (1911). Types: Zimbabwe, Chirinda Forest, *Swynnerton* 318 & 527 (K, syn.!, B, isosyn.!)

C. ituriense De Wild., Pl. Bequaert. 1: 17 (1921). Types: Zaire, Penghe, *Bequaert* 2513 (BR, syn.!) & Irumu, *Bequaert* 2938 (BR, syn.)
C. beniense De Wild., Pl. Bequaert. 1: 13 (1921). Type: Zaire, Mayolo, *Bequaert* 3981 (BR, holo.!)
C. limurense Rendle in J.B. 70: 158 (1932); Polhill in Journ. E. Afr. Nat. Hist. Soc. 24: 13 (1962). Type: Kenya, Kiambu District, Limuru, *Rendle* 636 (BM, holo.!)
C. elgonense Bullock in K.B. 1932: 503 (1932); Hanid in U.K.W.F.: 684 (1974). Type: Kenya, Elgon, *E.J. & C. Lugard* 629 (K, holo.!)
C. elatulum Poelln. in Portug. Acta Biol., sér. B, 1: 287 (1946). Type: Tanzania, Kigoma District, Uvinsa, W. Lugufu, *Peter* 46133 (B, holo.!)
C. glaucidulum Poelln. in Portug. Acta Biol., sér. B, 1: 293 (1946). Type: Tanzania, W. Usambara Mts., between Kwai and Gare, *Engler* 1220 (B, holo.!)
C. glaucidulum Poelln. var. *pauper* Poelln. in Portug. Acta Biol., sér. B, 1: 294 (1946). Type: Tanzania, Kilimanjaro, track towards Kibo, *Endlich* 692 (B, holo.!)
C. inopinum Poelln. in Portug. Acta Biol., sér. B, 1: 299 (1946). Type: Tanzania, E. Usambara Mts., Tengeni, *Peter* 23673 (B, holo.!) – form that tends to approach *C. gallabatense*
C. longum Poelln. in Portug. Acta Biol., sér. B, 1: 306 (1946). Type: Tanzania, Uluguru Mts. above Morogoro, Schlesien, *Peter* 32242 (B, holo.!)
C. macrophyllum (A. Rich.) Aschers. var. *angustifolium* Poelln. in Portug. Acta Biol., sér. B, 1: 310 (1946). Type: Tanzania, Usambara Mts., *Buchwald* 341 (B, holo.!)
C. magnum Poelln. in Portug. Acta Biol., sér. B, 1: 311 (1946). Type: Tanzania, Pangani District, Kalilanga outcrop, *Peter* 24522 (B, holo.!) – form that tends to approach *C. gallabatense*
C. nemorosum Poelln. in Portug. Acta Biol., sér. B, 1: 317 (1946). Type: Tanzania, Ulanga District, Mahenge, Mbangala, *Schlieben* 1780 (B, holo.!)
C. paludicolum Poelln. in Portug. Acta Biol., sér. B, 1: 320 (1946). Type: Tanzania, Mbulu District, Oldeani, *Kohl-Larsen* 369 (B, holo.!)
C. rugosum Poelln. in Portug. Acta Biol., sér. B, 1: 333 (1946). Type: Tanzania, North Pare, Kwa Muala Mts., *Peter* 14344 (B, holo.!)
C. turritum Poelln. in Portug. Acta Biol., sér. B, 1: 345 (1946). Type: Tanzania, Ukami, W. of Morogoro, *Peter* 46429 (B, holo.!, B, iso.!) – form that tends to approach *C. gallabatense*
C. usambarense Poelln. in Portug. Acta Biol., sér. B, 1: 347 (1946). Type: Tanzania, Lushoto District, E. Usambara Mts., Derema, *Scheffler* 237 (B, holo.!, B, iso.!)

NOTE. The species complex has up to recently been regarded as (at least) two different taxa, particularly delimited by reproduction systems: *C. comosum* with small plantlets originating at the inflorescence nodes (pseudovivipary) and *C. sparsiflorum* sensu lat. without such plantlets. The two reproductive traits are often displayed by plants belonging to the same populations, and cannot justify taxonomic delimitation. In addition to the variation in reproduction system, the species complex displays variation in several vegetative traits. When the species is found in drier, or more open habitats, the plants appear to respond by producing forms with ± erect peduncles and inflorescences with more condensed internodes. These forms have in East Africa often been identified to e.g. *C. kirkii*, *C. limurense* or *C. elgonense* (here reduced to synonymy). A particular robust form from the Uluguru Mountains and other montane areas has strap-shaped leaves drying blackish and rather large capsules, generally longer than wide. *C. longum* is the oldest name with type belonging here. In its strict sense *C. sparsiflorum* is, opposite to the forms above, distinctly petiolate, and appears to be most common in the western areas. It is probable that some of the mentioned forms deserve separate taxonomic rank, but until more detailed analyses are undertaken it is better to treat them as geographical variants or ecotypes without formal taxonomic recognition. More studies are greatly needed.

The *C. comosum* complex is furthermore not very well delimited from *C. filipendulum*, and particularly the montane '*longum*' form appears to be intermediate between the two taxa. *C. filipendulum* has, however, generally broader leaves, always distinctly petiolate, and most a often distinctly scabrid rhachis. The relation should be further analysed.

Pseudoviviparous forms of *C. comosum* are commonly used as ornamentals (sometimes with variegated leaves), either as garden plants in tropical to warm temperate areas or as indoor hanging-basket plants all over the world.

Chromosome number: 2n=14, based on countings from Uganda, Kenya and Tanzania, cf. Nordal et al. in Mitt. Inst. Allg. Bot. Hamburg 23 (1990).

47. **C. filipendulum** *Baker* in Trans. Linn. Soc., Bot., ser. 2, 1: 260 (1878) & in F.T.A. 7: 499 (1898). Type: Angola, Golungo Alto, *Welwitsch* 3776 (BM, holo.!, P, iso.!)

Plants 10–70 cm. high, immediately drying black when cut or when drying, often with a creeping stem. Rhizome usually elongated, horizontal to ascendent; roots thick, spongy expanding distally to spindle-shaped tubers 1–3 cm. long. Leaves rosulate, petiolate, petioles 10–20 cm. long, expanded and sheathing in lower parts; lamina broadly lanceolate, glabrous, never undulate, 12–50 cm. long, (3–)5–10 cm. wide, attenuate and apiculate. Peduncle terete, leafless (or with a few sterile bracts connected to the inflorescence), stout, 2–50 cm. long, glabrous. Inflorescence 5–30 cm. long, condensed or lax, usually unbranched, sometimes with 1–2 branches in lower part; rhachis ± scabrid; lower bracts 1.5–16 cm., ordinary flower bracts 1–2.5 cm., enveloping buds and pedicels. Pedicels articulated in the upper third and in forms with reduced peduncle and condensed inflorescence, just below the flower; whitish, particularly above the articulation, 6–8 mm. long in fruit, 3–4-nate at the lower nodes. Perianth whitish, sometimes with green tips, turning brownish immediately after anthesis, slightly urceolate; tepals patent above the constriction, 5.5–7 mm. long, 1.5–2.5 mm. wide, 3-veined. Stamens shorter to as long as the tepals; filaments fusiform, 3–5 mm. long, widest in upper half, scabrid; anthers 1.5–2.5 mm. long. Style slightly declinate. Capsule triquetrous in cross-section, emarginate, (5–)7–13 mm. long, generally distinctly longer than wide. Seeds saucer-shaped, several (± 8) per locule, 1.5–2 mm. across.

subsp. **filipendulum**

Peduncle longer than 10 cm. Inflorescence lax, usually with distinctly scabrid rhachis. Pedicels articulated at a distance below the flower.

UGANDA. Ankole District: Kashoya-Kitomi [Kasyoha-Kitomi] Forest Reserve, NE. of Kyambura R., 15 June 1994, *Poulsen* 515!; Mbale District: Bugisu [Bugishu] Forest, near Namatala R. on Bufumbo road, 3 Aug. 1953, *Norman* 228!; Mengo District: Kyiwaga [Kyewaga] Forest, 14 Sept. 1949, *Dawkins* 367!

KENYA. N. Kavirondo District: Kakamega Forest Station, 17 Apr. 1965, *Gillett* 16693! & Kakamega Forest, Kibiri Block, S. side of Yala R., 21 Jan. 1970, *Faden* 70/22!

TANZANIA. Bukoba District: Kaigi, May 1935, *Gillman* 280!; Lushoto District: Mt. Bomole, 3 Apr. 1950, *Verdcourt* 138!; Mpanda District: Kungwe-Mahali Peninsula, W. slopes of Musenabantu, 17 Aug. 1959, *Harley* 9369!

DISTR. **U** 2–4; **K** 5; **T** 1, 3, 4; widespread in the Guineo-Congolean rainforests from Ghana to Angola

HAB. In lower montane moist forests, on slopes and along streams; 950–1900 m.

subsp. **amaniense** (*Engl.*) *Nordal & A.D.Poulsen* stat. & comb. nov.

Peduncle shorter than 5 cm. Inflorescence condensed with glabrous rhachis. Pedicels articulated just below the flower.

KENYA. Kwale District: Jombo Mt., 8 Apr. 1968, *Magogo & Glover* 784!; Kilifi District: Marakaya, 6 May 1985, *Faden & Beentje* 85/48! & Marafa, 19 Nov. 1961, *Polhill & Paulo* 795!

TANZANIA. Pangani District: S. bank of Pangani R., between Hale and Makinyumbe, 1 July 1953, *Drummond & Hemsley* 3139!; Handeni District: near Kwedilumba, 23 Feb. 1987, *Pocs* 87025!; Morogoro District: Kimboza Forest Reserve, 30 Mar. 1983, *Mwasumbi, Rodgers & Hall* 12386!; Zanzibar, without precise locality, 1931, *Vaughan* 1329!

DISTR. **K** 7; **T** 3, 6; **Z**; not known elsewhere

HAB. In coastal or gallery evergreen dry to moist forests, on rocks; 30–650 m.

SYN. *C. amaniense* Engl. in E.J. 34: 157 (1904). Type: Tanzania, Lushoto District, Amani, *Engler* 2500 (B, holo.!)

NOTE. The relation between *C. filipendulum* and the *C. comosum* complex needs further study, and it may turn out that the species only represent extreme forms within the *C.*

comosum complex – see note under that species. The Asiatic *C. heynei* Baker in J.L.S. 15 (1876) has been regarded conspecific (Hanid pers. comm.), and this relationship needs to be further investigated.

The two subspecies are easily distinguished, although obviously closely related. Transitional forms between the two subspecies occur (e.g. *Drummond & Hemsley* 1456 from **T** 3). Most of the differences between them may be caused by simple regulations of internodal elongation. They deviate slightly also in the scabridity of the rhachis, subsp. *amaniense* being glabrous, but in this character also subsp. *filipendulum* from **T** 3 is less scabrid than forms found further west. Due to the almost apical pedicel-articulation of subsp. *amaniense*, it may be confused with *C. stenopetalum*. The latter is, however, a woodland species with different root system and is also easily delimited by the lack of a distinct petiole and narrower leaves.

For further notes, see also *C. macrophyllum*.

48. **C. lancifolium** *Baker* in Trans. Linn. Soc., Bot., ser. 2, 1: 260 (1878) & in F.T.A. 7: 498 (1898). Lectotype, chosen here: Angola, Pungo Andongo, *Welwitsch* 3772 (BM, lecto.!)

Plants 20–40 cm. high. Rhizome short; roots thick, spongy, with elongated distal tubers. Leaves drying black, rosulate, petiolate, obovate-lanceolate to oblong-lanceolate, truncate to cordate, 6–19 cm. long, 3–7 cm. wide; margins wavy with minute teeth. Peduncle terete, 8–20 cm. long, glabrous. Inflorescence paniculate to unbranched and lax; bracts supporting the branches (and floral bracts) relatively small, lower ones up to 15 mm. long. Pedicels articulated above the middle, 8–10 mm. long, 2–4-nate at the lower nodes. Perianth greenish white; tepals 4–6 mm. long, 3-veined. Stamens as long as the perianth; filaments papillate-rough, longer than the anthers. Style filiform, exserted. Capsule deeply triquetrous, 3–4 mm. long, 4–8 mm. wide. Seeds flat, 2.5–3 mm. in diameter.

UGANDA. Bunyoro District: Murchison Falls National Park, Rabongo Forest, 31 Aug. 1964, *H.E. Brown* 2123! & Budongo Forest, Kanyo-Pabidi block, 4 Feb. 1996, *Poulsen* 1184!

TANZANIA. Kigoma District: Uvinza, Malagarasi, Jan. 1926, *Peter* 35835b!; Tabora/Mpanda District: Ugalla Game Reserve, 1 Jan. 1974, *Nanai* 1837!; Ulanga District: Mahenge, Mar. 1932, *Schlieben* 1928!

DISTR. **U** 2; **T** 4, 6; widely distributed in the Guineo-Congolean parts of Africa, with extensions to the western parts of East Africa, Sudan, Angola and Zambia

HAB. In forest margins and closed woodland, on sandy rocky soils; 900–950 m.

SYN. *C. cordatum* Engl. in E.J. 15: 468 (1892); F.P.S. 3: 267 (1956). Type: Sudan, Niamniamland, *Schweinfurth* 173 (B, holo.!)

C. carsonii Baker in F.T.A. 7: 499 (1898). Type: Zambia, Urungu, Fwambo, *Carson* 29 (K, holo.!)

C. minutiflorum Poelln. in Portug. Acta Biol., sér. B, 1: 314 (1946). Type: Tanzania, Kigoma District, Uvinza, Malagarasi, *Peter* 35835b (B, holo.!)

C. nebulosum Poelln. in Portug. Acta Biol., sér. B, 1: 317 (1946). Type: Tanzania, Ulanga District, Mahenge, *Schlieben* 1928 (B, holo.!)

C. peteri Poelln. in Portug. Acta Biol., sér. B, 1: 324 (1946). Type: Tanzania, Kigoma District, W. of Uvinza, *Peter* 36391 (B, holo.!)

NOTE. *C. lancifolium* closely resembles the widespread tropical species, *C. orchidastrum*. The latter has attenuate rather than cordate leaf-bases, and often exhibits rather large and leafy bracts in the inflorescence. Chromosome number: 2n=14, based on countings from Tanzania, cf. Nordal et al. in Mitt. Inst. Allg. Bot. Hamburg 23 (1990).

49. **C. holstii** *Engl.*, P.O.A. C: 140 (1895); Baker in F.T.A. 7: 497 (1898). Type: Tanzania, Tanga District, Usambara Mts., Amboni, *Holst* 2674 (B, holo.!, K, iso.!)

Small plants up to 15 cm. Rhizome short; roots rather thin with tubers on short lateral branches. Leaves drying yellowish brown, rosulate, sheathing, with distinctly canaliculate petiole, 5–10 cm. long; lamina forming a distinct angle somewhere between 90° and 180° with the petiole so that it is parallel with the soil surface, 10–20 cm. long, 1–3 cm. wide, papillose below at least along the midvein, distinct teeth along the margin, leaf-base attenuate, apex obtuse to acute. Peduncle slender, 5–20 cm. long, scabrid to papillate. Inflorescence a lax raceme, 8–20 cm. long, simple or slightly branched at the base with ± pubescent rhachis; floral bracts papillose, 7–18 mm., obovoid, acute. Pedicels articulated in lower half, 5–6 mm. long in flower, up to 12 mm. in fruit, scabrid to pubescent, 2–4-nate at the lower nodes. Perianth white; tepals patent, 5–8 mm. long, 1.5–3 mm. wide, 3-veined. Filaments fusiform, 2–4 mm. long, with scale-like papillae; anthers ± 2 mm. Capsule trigonous, ± 5–7 mm. long, 5–9 mm. wide, with truncate base. Seeds 2 per locule, saucer-shaped, relatively large, 3–3.5 mm. in diameter.

KENYA. Kwale District: Mwachi Forest Reserve, 17 May 1990, *Robertson & Luke* 6240!
TANZANIA. Tanga District: 8 km. S. of Ngomeni, 29 July 1953, *Drummond & Hemsley* 3504! & Kange Gorge, 19 Apr. 1950, *Faulkner* 1850!; Mpwapwa, 22 Feb. 1934, *Hornby* 630!
DISTR. **K** 7; **T** 3–6; not known elsewhere
HAB. On rocky hillsides in dense dry evergreen lowland or riverine forest or *Brachystegia* woodland, also in thick succulent/thorn scrub, 0–± 1000 m.

SYN. *C. hoffmannii* Engl. in E.J. 34: 158 (1904). Type: Tanzania, Handeni District, Useguha, cult. Berlin, *Hoffmann & Stolz* (B, holo.!)
C. holstii Engl. var. *glabrum* Poelln. in Portug. Acta Biol., sér. B, 1: 298 (1946). Type: Tanzania, E. Usambara Mts., Amani, *Peter* 16230 (B, holo.!, B, iso.!)
C. pulverulentum Poelln. in Portug. Acta Biol., sér. B, 1: 329 (1946). Type: Tanzania, Handeni District, Useguha, between Sindeni and Handeni, *Peter* 40578 (B, holo.!)
C. uvinsense Poelln. in Portug. Acta Biol., sér. B, 1: 348 (1946). Type: Tanzania, Kigoma District, E. of Uvinza, *Peter* 36263 (B, holo.!)

NOTE. *C. holstii* has often been identified as the Guineo-Congolean rain-forest species, *C. alismifolium* Baker, to which it is obviously related (cf. *C. sp. B*). They differ, however, by several traits: *C. alismifolium* has thicker, tomentose roots, with tubers as thickenings along the main axes, not on lateral branches; the leaves display less dense venation in that species; it has a shorter more erect peduncle, only up to 5 cm. long; the tepals are widest in the upper part, and not ± parallel-sided as in *C. holstii*; and in particular, the capsules are distinctly stipitate in *C. alismifolium*, but truncate, without any sign of a stipe, in *C. holstii*. The two taxa might represent vicariant subspecies of *C. alismifolium*; more material and more thorough analyses are needed.

50. **C. tenerrimum** *Poelln.* in Portug. Acta Biol., sér. B, 1: 340 (1946). Type: Tanzania, Lushoto District, E. Usambara Mts., Makumba, *Peter* 16326 (B, holo.!)

Slender grass-like plants up to 25 cm. Rhizome short; roots rather thin with tubers along the main axes. Leaves drying yellowish brown, rosulate, linear, without a distinct petiole, 12–25 cm. long, up to 0.5 cm. wide, apex acute, glabrous except for cilia along the margin, particularly in the apical parts. Peduncle slender, ± 5 cm. long, glabrous. Inflorescence a lax raceme, 3–10 cm. long, simple or slightly branched at the base, with glabrous rhachis; floral bracts obovoid, 3–10 mm. long, acute, toothed apically. Pedicels articulated in lower half, 7–8 mm. long in fruit, glabrous, 1–2-nate at the lower nodes. Perianth white; tepals 4–5 mm. long, 1.5–2 mm. wide, 3-veined. Filaments fusiform and papillate, 3–3.5 mm. long; anthers ± 1.5 mm. long. Capsule trigonous, 6–9 mm. long, 7–9 mm. wide, widest near the apex, attenuate and slightly stipitate, at base. Seeds few per locule, saucer-shaped, up to 3 mm. in diameter.

KENYA. Kwale District: Buda Mafisini Forest, 15 Aug. 1953, *Drummond & Hemsley* 3794!; Kilifi District: Arabuko-Sokoke Forest, Jilore [Jilori], 25 Nov. 1961, *Polhill & Paulo* 851! & N. of Sokoke Forest Station, 8 June 1973, *Musyaki & Hansen* 994!
TANZANIA. Tanga District: 8 km. SE. of Ngomeni, 29 July 1953, *Drummond & Hemsley* 3507!; Pangani District: Msubugwe Forest, 26 Aug. 1955, *Tanner* 2113!
DISTR. **K** 7; **T** 3; not known elsewhere
HAB. Dry lowland forest and tall shrub thickets, on red sandy soils; 70–150 m.

SYN. [*C. laxum* sensu Baker in F.T.A. 7: 503 (1898) pro parte, *non* R. Br.]
C. sp. near *laxum* sensu Polhill in Journ. E. Afr. Nat. Hist. Soc. 24: 14 (1962)

NOTE. A dwarfed coastal forest plant that often has often been identified as *C. laxum* R. Br. The latter is, however, described and typified by Australian plants to which the relation is unclear (see also note under *C. inconspicuum*). The stipitate ovary might indicate a relationship to the guineo-Congolean forest species *C. alismifolium* Baker (see also note under *C. sp. A*).

51. **C. sp. A**

Robust plants up to 60 cm. Rhizome short, carrying medium thick roots with elongated distal tubers. Leaves ± 60 cm. long, 10 cm. wide, widest above the middle, gradually narrowing towards the base, pubescent, ciliate along margin and on veins. Peduncle bracteate, 15–30 cm. long. Inflorescence paniculate, with 2–4 flowers at each node. Pedicels ± 4 mm. long, articulated in upper half. Flowers not known. Fruits triquetrous, 4–5 mm. long, 4–5 mm. wide. Seeds saucer-shaped, ± 2 mm. in diameter.

UGANDA. Toro District: Bwamba County, 1–2 km. S. of Sempaya, 23 Sept. 1969, *Lye* 4293! & Bwamba, 22 July 1958, *A.S. Thomas* 2311!
DISTR. **U** 2; not known elsewhere
HAB. On forest floor; 700 m.

NOTE. This is obviously an undescribed species. It connects to *C. andongense* by the bracteate peduncle and the general habit. It differs, however, by the pubescent leaves and the rather small capsules. More, and particularly flowering, material is needed.

52. **C. sp. B**

Small plants up to 10 cm., with creeping stem. Rhizome short; roots thick with distal tubers. Long membranous cataphylls; ordinary leaves drying green, rosulate, glabrous, sheathing, with distinctly canaliculate petiole, 5–7 cm. long; lamina forming a distinct angle somewhere between 90° and 180° with the petiole so that it is parallel with the soil surface, ± 7 cm. long, 4 cm. wide, base truncate. Peduncle ± 3 cm. long. Inflorescence condensed, 1–2 cm. long, hiding below the leaves; bracts membranous, up to 5 mm. long, supporting one flower at each node; pedicel 2–3 mm. long in fruit, articulated above middle. Flowers not known. Fruits green, 4–5 mm. long, 7–8 mm. wide, ± truncate at the base. Seeds saucer-shaped, large, up to 4 mm. in diameter, only one per locule.

UGANDA. Bunyoro District: Budongo Forest Reserve, close to the Sonso R., 2 Oct. 1995, *Poulsen* 975!
DISTR. **U** 2; E. Zaire
HAB. Forest; 1000 m.

NOTE. This species is obviously closely related to the Guineo-Congolean *C. alismifolium* Baker. It differs, however, by the very short peduncle and hidden inflorescence, and also by the base of the capsule, being stipitate in the latter. More, and particularly flowering, material is needed.

UNCERTAIN NAMES

The types of these names have not been found and the descriptions are too general to decide on their taxonomic position.

Anthericum ulugurense Engl. in E.J. 28: 360 (1900). Type: Tanzania, Morogoro District, Uluguru Mts., Mbakana, *Goetze* 353 (B, holo.) – might according to the description belong in *C. cameronii*

Anthericum papillosum Engl. in E.J. 28: 360 (1900). Type: Tanzania, Morogoro District, Uluguru Mts., Mbakana, *Goetze* 353a (B, holo.)

Anthericum princeae Engl. & K. Krause in E.J. 45: 129 (1910). Type: Tanzania, Iringa, Ubeno, *Prince* (B, lecto.) – might according to the description belong in *C. cameronii*

Anthericum jaegeri Engl. & K. Krause in E.J. 45: 130 (1910). Type: Tanzania, Masai District, E. of Ikoma, Lamuniane Hills, *Jaeger* 353 (B, holo.) – might according to the description belong in *C. cameronii*

Anthericum oehleri Engl. & K. Krause in E.J. 45: 130 (1910). Type: Tanzania, Masai District, E. of Ikoma, Lamuniane Hills, *Jaeger* 354 (B, holo.) – might according to the description belong in *C. cameronii*

Anthericum saxicolum Poelln. in F.R. 51: 32 (1942). Type: Tanzania, Lushoto District, Usambara Mts., Kwai, *Engler* 1232 (B, holo.)

Anthericum dilatatum Poelln. in F.R. 51: 68 (1942). Type: Tanzania, Iringa District, Uhehe, Uzungwa [Utschungwe] Mts., *Prince* (B, holo.) – might according to the description belong in *C. cameronii*

Anthericum uyuiense Rendle var. *latifolium* Poelln. in F.R. 51: 75 (1942). Type: Tanzania, Pangani District, Useguha, Hale, *Peter* 24337 (B, holo.) – might according to the description belong in *C. cameronii*

Anthericum depauperatum Poelln. in F.R. 51: 77 (1942). Type: Tanzania, Morogoro District, Uluguru Mts., Magogoni to Ruvu [Ruwu], *Brehmer* 765 (B, holo.)

Anthericum grantii Baker var. *longipedicellatum* Poelln. in F.R. 51: 117 (1942). Type: Tanzania, Tabora District, Igonda [Gonda], *Boehm* 31 pro parte (B, holo.) – might according to the description belong in *C. cameronii*

Anthericum amplexifolium Poelln. in F.R. 51: 120 (1942). Type: Tanzania, Kigoma District, Uvinza to Malagarasi, *von Trotha* 68 (B, holo.) – might according to the description belong in *C. blepharophyllum*

Anthericum ramiferum Poelln. in F.R. 51: 129 (1942). Type: Tanzania, Tabora District, Ngulu, from Malongwe W. to Nyahua, *Peter* 34697 (B, holo.)

Anthericum princeae Engl. & K. Krause var. *brevibracteatum* Poelln. in F.R. 51: 134 (1942). Type: Tanzania, Ufipa District, Msambia [Msamvia], *Münzner* 7203 (B, holo.)

Excluded Names

Anthericum oatesii Baker in J.B.: 324 (1878) & in F.T.A. 7: 491 (1898) – belongs in *Trachyandra* (Asphodelaceae)

Anthericum saltii Baker in J.L.S. 15: 309 (1898) & in F.T.A. 7: 492 (1898) – belongs in *Trachyandra* (Asphodelaceae)

Chlorophytum pilosum Dammer in E.J. 48: 364 (1912) – belongs in *Trachyandra* (Asphodelaceae)

Anthericum kassneri Poelln. in F.R. 50: 322 (1941) – belongs in *Trachyandra* (Asphodelaceae)

Anthericum albovaginatum Poelln. in F.R. 51: 27 (1942) – belongs in *Trachyandra* (Asphodelaceae)

Anthericum blepharophyllum Poelln. in F.R. 51: 135 (1942) – belongs in *Trachyandra* (Asphodelaceae)

Chlorophytum latifilimentatum Poelln. in Portug. Acta Biol., sér. B, 1: 357 (1946) – belongs in *Drimiopsis* (Hyacinthaceae)

INDEX TO ANTHERICACEAE

GEOGRAPHICAL DIVISIONS OF THE FLORA

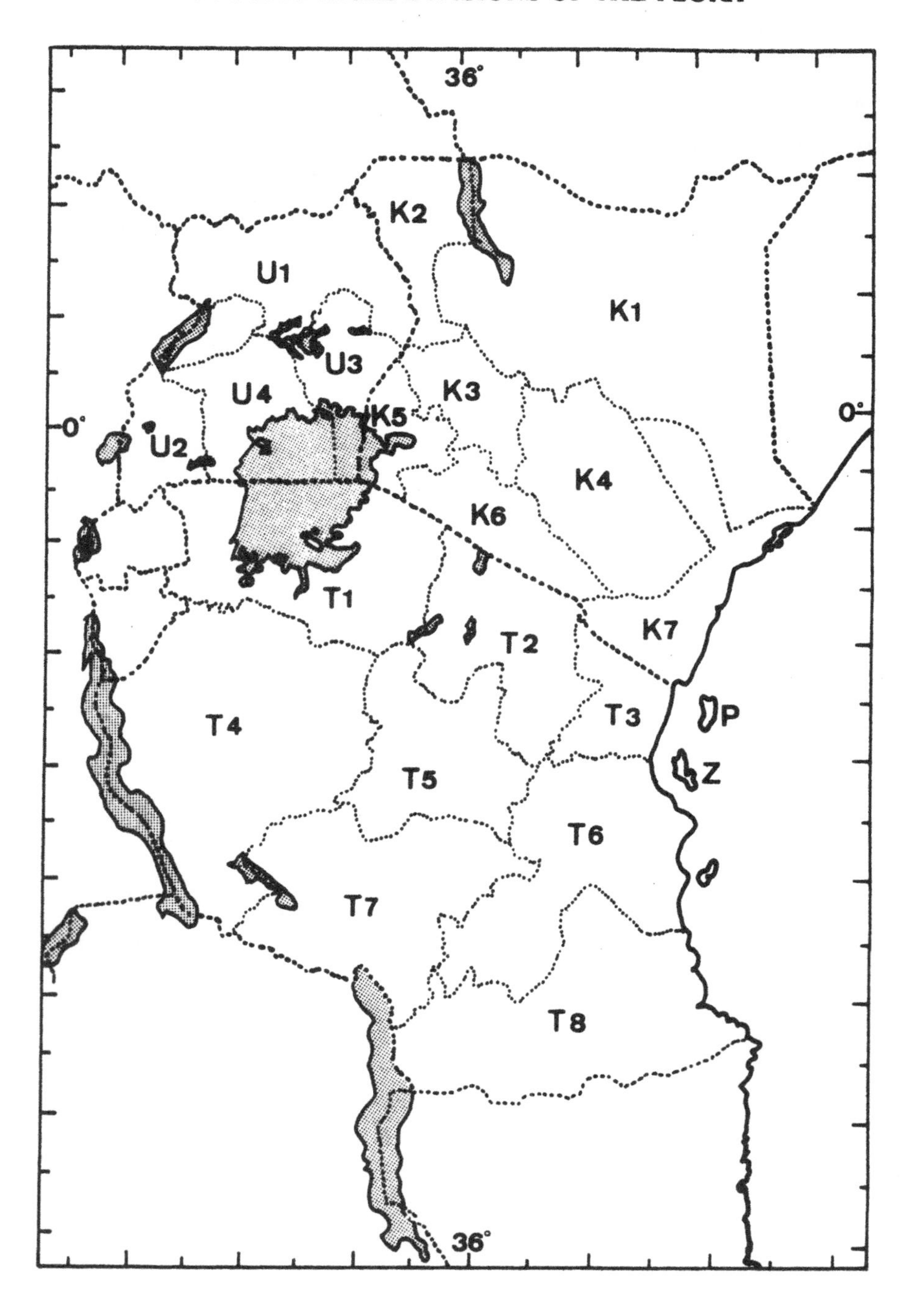

For Product Safety Concerns and Information please contact our EU representative GPSR@taylorandfrancis.com Taylor & Francis Verlag GmbH, Kaufingerstraße 24, 80331 München, Germany

Printed and bound by CPI Group (UK) Ltd, Croydon, CR0 4YY
01/05/2025
01858466-0001